BACK-OF-THE-ENVELOPE QUANTUM MECHANICS

With Extensions to Many-Body Systems, Integrable PDEs, and Rare and Exotic Methods

Second Edition

BACK-OF-THE-ENVELOPE QUANTUM MECHANICS

With Extensions to Many-Body Systems, Integrable PDEs, and Rare and Exotic Methods

Second Edition

Maxim Olshanii

University of Massachusetts Boston, USA

NEW JERSEY · LONDON · SINGAPORE · BEIJING · SHANGHAI · HONG KONG · TAIPEI · CHENNAI · TOKYO

Published by

World Scientific Publishing Co. Pte. Ltd.

5 Toh Tuck Link, Singapore 596224

USA office: 27 Warren Street, Suite 401-402, Hackensack, NJ 07601

UK office: 57 Shelton Street, Covent Garden, London WC2H 9HE

Library of Congress Control Number: 2023952220

British Library Cataloguing-in-Publication Data
A catalogue record for this book is available from the British Library.

BACK-OF-THE-ENVELOPE QUANTUM MECHANICS
With Extensions to Many-Body Systems, Integrable PDEs, and Rare and Exotic Methods
Second Edition

ISBN 978-981-12-8637-7 (hardcover)
ISBN 978-981-12-8638-4 (ebook for institutions)
ISBN 978-981-12-8639-1 (ebook for individuals)

For any available supplementary material, please visit
https://www.worldscientific.com/worldscibooks/10.1142/13680#t=suppl

Typeset by Stallion Press
Email: enquiries@stallionpress.com

Printed in Singapore

To Dimitri, Mark, Laura, Sophia, Milena, and to my Father

Preface

As I now realize it, this book was uniquely inspired by Professor Krainov's course on qualitative methods in physical kinetics that I attended at the Moscow Engineering Physics Institute (National Research Nuclear University MEPhI nowadays) thirty years ago. As we students would learn in a more rigorous class to follow, in physical kinetics, even the most basic results require laborious multi-page derivations. But Krainov's course and his book published later by the American Institute of Physics taught us that if one is *not* interested in the exact values of prefactors, then ten pages of calculations can be replaced by two short lines on the *back of an envelope*; and in some cases, even a postal stamp would suffice.

The book you are about to read is based on the problems assigned in a graduate course in quantum mechanics that I have been teaching at the University of Massachusetts Boston for many years. Similarly to the physical kinetics classes I attended at the MEPhI, the discussion on any new topic in my class would invariably start from a series of qualitative problems. When I realized I had more than fifty of them, I decided to assemble them in a book.

In this book, I clearly *distinguish between the dimensional and the order-of-magnitude estimates*. Dimensional analysis is a powerful method to analyze new unexplored equations, but it fails when there are too many dimensionless parameters involved. In an order-of-magnitude estimate— a calculation where all angles are 90°, all numbers are unity, and all integrals are just "height times width"—one needs to understand the physics behind the process really well; as a reward, the method is nearly universal.

Approximately half of the book is devoted to the estimates based on either semi-classical approximation or on perturbation theory expansions in *elementary quantum mechanics*. Thanks to a reduced number of independent dimensionful parameters in the domains of applicability of these theories, both dimensional and order-of-magnitude approaches are ideally suited there.

A sequence of variational problems is also included. The breadth and elegance of variational reasoning makes it a valuable tool in a preliminary analysis of a problem; determination of the parity of the ground state in a well is a good example. Furthermore, even the quantitative results obtained from simple one-parametric variational ansatzes still fit on an envelope.

Similarly, I could not resist including several powerful results produced by applying the Hellmann-Feynman theorem to *integrable many-body quantum systems*. Unlike other methods considered, it produces exact answers; those can also be obtained in a few lines.

The *integrable partial differential equations* serve as an example of a field where there are no innate measurement units, and yet dimensional analysis can be deployed; the dependence of the size of a Koteweg-de Vries soliton on its speed is a typical application of the method.

This book contains both solved problems and exercises. The order of the solved problems is important: the sequence gradually prepares the reader for the problems without solutions. Minimal theoretical background is provided as well. Several lesser known theoretical facts are attached to the respective "Background" sections as "Problems linked to the 'Background' ". Various approximate and qualitative methods are compared in three case studies: of a hybrid, harmonic-quartic oscillator, of a "halved" harmonic oscillator, and of a gravitational well.

This book would not have been possible without input from all the students I have taught in my quantum mechanics courses at UMass Boston and at the University of Southern California before it. Special thanks to Vladimir Pavlovich Krainov for introducing me to qualitative methods, first as a professor and, later on, as my first research project adviser. Further interactions with my mentors, Vladimir Minogin and Yvan Castin, inspired many new problems for the book and shaped its structure.

A good half of this book was compiled during quiet Mediterranean nights, profiting from the free internet in the lobby of the Galil Hotel in Netanya, Israel. Many thanks to its stuff for the cookies they were incessantly feeding me throughout those nights.

This is an appropriate place to thank my friends—Vincent Lorent, Lana Jitomirskaya, Vanja Dunjko, Lena Dotsenko, and Paul Gron—for standing by my side in good and in bad times.

I am immensely grateful to Zaijong Hwang and Vanja Dunjko for a thorough critical reading of the manuscript.

Finally, I would like to thank my wife Milena Gueorguieva for correcting commas, articles, and awkward sentences and my son Mark Olchanyi for producing the cover art.

Maxim Olshanii
Boston, Massachusetts
January 14, 2013

Second Edition: an update. Misprints of the First Edition have been corrected. I am especially grateful to Robert Barr and Mary Fries for spotting most of them.

I am indebted to all the students in my *Estimates in Physics* class, at UMass Boston: they were the first readers and the first critics of almost all the new problems.

Again, I am immensely grateful to Zaijong Hwang and Vanja Dunjko for a thorough critical reading of the manuscript, both editions.

The Second Edition features two new Chapters: Chapter 2, *A dimensional estimate for Planck's energy: a Case Study* and Chapter 12, *Rare and exotic methods in elementary quantum mechanics and beyond*.

Chapter 2 complements the Chapter 1; the combination of the two serves as an introduction to varios types of estimates. The new Chapter is also intended to demonstrate that unlike for any other method in physics, dimensional analysis does not require any knowledge in the field to which the method is applied.

The First Edition of the book was devoted to the *generally applicable* shortcuts that allow one to bypass solving differential equations, computing complicated integrals, and diagonalizing infinite-size matrices. These shortcuts included: order-of-magnitude estimates, dimensional analysis, and variational methods. The Chapter 12, which makes its first appearance in the Second Edition, complements this list: it focusses on the methods whose *applicability is limited* to a narrow class of problems.

Maxim Olshanii
Boston, Massachusetts
March 15, 2023

Contents

Some notations

$[a]$	Units in which an observable a is measured
$[\mathcal{L}], [\mathcal{T}], [\mathcal{M}], [\mathcal{E}], \ldots$	Units of length, time, mass, energy, $\ldots$
$\mathcal{L}, \mathcal{T}, \mathcal{M}, \mathcal{E}, \ldots$	Length scale, time scale , mass scale , energy scale, $\ldots$
$P_1, P_2, \ldots$	Independent dimensionless parameters of a given problem
$\mathcal{A}[\psi(\cdot)]$	A functional $\mathcal{A}$ acting on a wavefunction $\psi(x)$
$\mathfrak{V}$	A variational space $\mathfrak{V}$

Chapter 1

Ground State Energy of a Hybrid Harmonic-Quartic Oscillator: a Case Study

Introduction

Consider the Schrödinger equation for a one-dimensional particle moving in a combination of harmonic potential of frequency ω and a quartic potential of strength β:

$$-\frac{\hbar^2}{2m}\frac{\partial^2}{\partial x^2}\psi(x) + \frac{m\omega^2}{2}x^2\psi(x) + \beta x^4\psi(x) = E\psi(x)\,, \qquad (1.1)$$

where m is the particle's mass. We will be mainly interested in determining the ground state energy. The Eq. (1.1) does not allow for an exact solution. However, the major features of the dependence of the ground state energy on the system parameters can be determined via elementary methods, such as dimensional analysis, order-of-magnitude estimates, and simple variational bounds. The goal of this Chapter is to illustrate the application of these methods using the ground state problem (1.1) as an example.

1.1 Solved problems

1.1.1 *Dimensional analysis and why it fails in this case*

The assignment is: perform dimensional analysis of the problem and show that from a dimensional point of view the problem is underdetermined: no estimate for the ground state energy can be produced. However, some information about the structure of the expression for the ground state energy can still be extracted, on purely dimensional grounds.

Solution: The dimensional procedure for finding the ground state energy $E_{\text{g.s.}}$ (or assessing the impossibility of a complete dimensional solution) is as follows:

– Begin by identifying the *principal units* of measurement for the problem, i.e. the minimal set of units sufficient to describe all input parameters of the problem. For stationary problems in quantum mechanics, the units of length, $[\mathcal{L}]$, and energy, $[\mathcal{E}]$, have been proven to provide a handy set;

– Identify the *input parameters* and units used to measure them;

– Determine the *maximal set of independent dimensionless parameters*: the set will include only the parameters that are generally either much greater or much less than unity. These include both the dimensionless parameters present in the problem a priori (such as the quantum number n), and the dimensionless combinations of the dimensionful input parameters. If the set is empty, the unknown quantities can be determined almost completely, i.e. up to a numerical prefactor of the order of unity. If some dimensionless parameters are present, the class of possible relationships between the unknowns and the input parameters can be narrowed down, but the order of magnitude of the unknown quantities can not be determined.

– For each of the principal units, choose a *scale*: a combination of the input parameters measured using the unit in question;

– Express the unknown quantities as a multi-power-law of principal scales, times an arbitrary function of all dimensionless parameters, if any. If no dimensionless parameters are present, the arbitrary function is replaced by an arbitrary constant, presumed to be of the order of unity.

In our case, the above procedure gives:

– *The principal units—the units of length and the units of energy*:

$$[\mathcal{L}], \quad [\mathcal{E}];$$

– *The input parameters and their units*:

$$[\eta] = [\mathcal{L}]^2 [\mathcal{E}]$$
$$[\Upsilon] = [\mathcal{L}]^{-2} [\mathcal{E}]$$
$$[\beta] = [\mathcal{L}]^{-4} [\mathcal{E}],$$

where $\eta \equiv \hbar^2/m$, and $\Upsilon \equiv m\omega^2$;

– *The set of independent dimensionless parameters* $=$

$$\left\{ P_1 \equiv \frac{\hbar\beta}{m^2\omega^3} \right\}. \tag{1.2}$$

It is represented by a single element. Let us prove that. First, there are no a priori dimensionless parameters in this problem. Assume now that P is a dimensionless parameter that is derived from the dimensionful input parameters. It must be represented as a multi-power law of the input parameters:

$$P = \text{const} \times \eta^{\xi_1} \Upsilon^{\xi_2} \beta^{\xi_3} .$$

Units for P are given by

$$
\begin{aligned}
[P] &= [\eta]^{\xi_1} [\Upsilon]^{\xi_2} [\beta]^{\xi_3} \\
&= \left([\mathcal{L}]^2 [\mathcal{E}] \right)^{\xi_1} \left([\mathcal{L}]^{-2} [\mathcal{E}] \right)^{\xi_2} \left([\mathcal{L}]^{-4} [\mathcal{E}] \right)^{\xi_3} \\
&= [\mathcal{L}]^{2\xi_1 - 2\xi_2 - 4\xi_3} [\mathcal{E}]^{\xi_1 + \xi_2 + \xi_3} .
\end{aligned}
$$

On the other hand, P is supposed to be dimensionless:

$$[P] = [\mathcal{L}]^0 [\mathcal{E}]^0 .$$

Thus, the powers ξ_1, ξ_2, ξ_3 must obey the following system of linear homogeneous algebraic equations:

$$\hat{M} \cdot \begin{pmatrix} \xi_1 \\ \xi_2 \\ \xi_3 \end{pmatrix} = \begin{pmatrix} 0 \\ 0 \end{pmatrix}, \qquad (1.3)$$

where

$$\hat{M} = \begin{pmatrix} 1 & 1 & 1 \\ 2 & -2 & -4 \end{pmatrix}. \qquad (1.4)$$

The number of independent dimensionless parameters will be given by

$$
\begin{aligned}
&(\# \text{ of independent dimensionless parameters}) \\
&= \underbrace{(\# \text{ of independent } a\text{-}priori\text{-dimensionless parameters})}_{0} \\
&+ \underbrace{(\# \text{ of independent dimensionful parameters})}_{3} \\
&\underbrace{- \text{rank}(\hat{M})}_{2} \\
&= 1. \qquad (1.5)
\end{aligned}
$$

The dimensionless parameter P_1 can be found by solving the system (1.3). It gives

$$\begin{pmatrix} \xi_1 \\ \xi_2 \\ \xi_3 \end{pmatrix} = \text{const} \times \begin{pmatrix} \frac{1}{2} \\ -\frac{3}{2} \\ 1 \end{pmatrix},$$

leading to

$$P_1 = \text{const} \times \frac{\beta\sqrt{\eta}}{\Upsilon^{3/2}}$$

$$= \text{const} \times \frac{\beta\hbar}{m^2\omega^3} \; ;$$

– *The principal scales—the length scale and the energy scale, examples of*:

$$\mathcal{L} = \sqrt{\frac{\hbar}{m\omega}}$$

$$\mathcal{E} = \hbar\omega \; .$$

The principal scales above are defined as examples of observables measured in principal units, $[\mathcal{L}]$ and $[\mathcal{E}]$ in our case. To derive the above expression for the length scale, let us represent this scale as

$$\mathcal{L} = \text{const} \times \eta^{\nu_1} \Upsilon^{\nu_2} \beta^{\nu_3} \; .$$

The corresponding units are related as

$$[\mathcal{L}] = [\eta]^{\nu_1} [\Upsilon]^{\nu_2} [\beta]^{\nu_3}$$

$$= \left([\mathcal{L}]^2 [\mathcal{E}]\right)^{\nu_1} \left([\mathcal{L}]^{-2} [\mathcal{E}]\right)^{\nu_2} \left([\mathcal{L}]^{-4} [\mathcal{E}]\right)^{\nu_3}$$

$$= [\mathcal{L}]^{2\nu_1 - 2\nu_2 - 4\nu_3} [\mathcal{E}]^{\nu_1 + \nu_2 + \nu_3} \; .$$

The powers ν_1, ν_2, ν_3 obviously obey a system of linear inhomogeneous algebraic equations given by

$$\hat{M} \begin{pmatrix} \nu_1 \\ \nu_2 \\ \nu_3 \end{pmatrix} = \begin{pmatrix} 0 \\ 1 \end{pmatrix} , \tag{1.6}$$

where M is given by the expression (1.4). Any particular solution of Eq. 1.6 (and in this particular case we have a one-dimensional family of them) can be chosen to represent a length scale; this choice is a matter of convenience. We choose the scale associated uniquely with the harmonic oscillator,

$$\begin{pmatrix} \nu_1 \\ \nu_2 \\ \nu_3 \end{pmatrix} = \begin{pmatrix} \frac{1}{4} \\ -\frac{1}{4} \\ 0 \end{pmatrix} ,$$

or, for example,

$$\mathcal{L} = \left(\frac{\beta\eta}{\Upsilon}\right)^{1/4}$$

$$= \sqrt{\frac{\hbar}{m\omega}} \; .$$

The energy scale $\mathcal{E}$, given by the expression (1.6), can be obtained the same way. The only difference is the right hand side of equation (1.6): it should read $\begin{pmatrix} 1 \\ 0 \end{pmatrix}$;

– *Solution for the unknown*:

$$[E_{\text{g.s.}}] = [\mathcal{E}] \Rightarrow E_{\text{g.s.}} = \Phi(P_1) \times \mathcal{E} = \Phi(\frac{\hbar\beta}{m^2\omega^3}) \times \hbar\omega . \quad (1.7)$$

where $\Phi(P)$ is an arbitrary function.

To summarize,

$$
\boxed{
\begin{array}{c}
E_{\text{g.s.}} = \Phi(\dfrac{\hbar\beta}{m^2\omega^3}) \times \hbar\omega \\[2mm]
\Phi(\cdot) = \text{any function}
\end{array}
}
$$

This solution does narrow the class of possible expressions for the ground state energy, but does not allow one to determine it, not even its order of magnitude.

If needed, analogous expressions for other observables can be readily obtained. For the observables measured in combinations of principal units only, one should combine the principal scales to form a scale for the observable of interest. The full dimensional prediction for this observable will be given, as before, as a product of an arbitrary function of all dimensionless parameters and the scale. For example, the r.m.s. (root mean square) force acting on our particle in the ground state will be given by

$$F_{\text{g.s.}} = \Phi_2(P_1) \times \mathcal{F} = \Phi_2(\frac{\hbar\beta}{m^2\omega^3}) \times \sqrt{m\hbar\omega^3} ,$$

where

$$\mathcal{F} \equiv \frac{\mathcal{E}}{\mathcal{L}} = \sqrt{m\hbar\omega^3}$$

is the force scale, and $\Phi_2(P)$ is another arbitrary function.

For the observables measured in units that do not belong to the principal set (the minimal set of units to describe all input parameters), other scales must be invented if needed. For example, the inverse harmonic frequency, $1/\omega$, provides a useful time scale:

$$\mathcal{T} = \frac{1}{\omega} .$$

In mechanics problems, both classical and quantum, no more than three independent scales are ever necessary. For example, the r.m.s. (root mean square) ground state velocity is given by

$$v_{\text{g.s.}} = \Phi_3(P_1) \times \mathcal{V} = \Phi_3\left(\frac{\hbar\beta}{m^2\omega^3}\right) \times \sqrt{\hbar\omega/m}\,,$$

with

$$\mathcal{V} \equiv \frac{\mathcal{L}}{\mathcal{T}} = \sqrt{\hbar\omega/m}$$

being the velocity scale.

1.1.1.1   *Side comment: dimensional analysis and approximations*

Often, when an exact theory is replaced by an approximate one, the number of independent dimensionful parameters decreases, thus shifting the counting (1.5) in favor of accurate up-to-a-prefactor dimensinal predictions. In what follows, we will encounter several examples of such reduction. For example, in semi-classical theory, considered in Chapter 3, the Plank constant $\hbar$ and the level index n fuse into a single entity, the classical canonical action $\tilde{\hbar}$:

This reduction is a consequence of the absence of the quantization of action in classical mechanics. Another example is the Thomas-Fermi theory, Chapter 10. There, the number of electrons Z and the electron charge $|e|$ unite and form a total charge Q, never appearing separately:

The physical reason for such a merger is that in mean-field theories—such as the Thomas-Fermi theory—the number of electrons is no longer quantized, and a continuous field of electron number density $n(\vec{r})$ is used instead of the individual electron positions. A similar reduction happens when solutions of a many-body Schrödinger equation for an ensemble of bosonic particles are approximated using solutions of a one-body nonlinear Schrödinger equation (See Problem 11.1.3).

Yet another example of a reduction of the number of input parameters under an approximation is provided by perturbation theory (Chapter 5). Here, in the n-th order of perturbation theory, the power of a dimensionful prefactor in front of the perturbation is fixed to n, effectively removing the prefactor from the list of independent parameters. See, specifically, Problems 5.1.8 and 5.1.9.

1.1.1.2   *Side comment: how to recast input equations in a dimensionless form*

Prior to involved analytical or numerical calculations, equations are often expressed in a "dimensionless form", the advantage being a reduced number of parameters and an absence of numbers that are too large or too small.

The recipe is as follows:

> - Several input parameters, as many as there are principal units, are replaced by unity;
> - The remaining input parameters retain their notations, but their numerical dimensionful values are replaced by dimensionless numbers given by the ratios between the original values of the parameters and the corresponding "scales";
> - Likewise, in the end the numerical values of the answers are multiplied by the corresponding scales.

A well-defined formal procedure is hidden behind. It consists of two elements. (a) The parameters to be set to unity are chosen as principal scales; (b) these scales and their multi-power combinations are used as new units of measurement.

In our example, we can choose $[\eta]$ and $[\Upsilon]$ as the "principal units" and η and Υ themselves as the "principal scales". Obviously, in this system of units, η and Υ assume unit values. Also, for moderate values of the remaining dimensionless parameters, all answers we get become of the order of unity.

Conventionally, this system of units is denoted as

$$\frac{\hbar^2}{m} = m\omega^2 = 1\,,$$

or even more often

$$\hbar = m = \omega = 1\,. \tag{1.8}$$

The appearance of three scales is not an accident. The recipe (1.8) (a) does not lead to any ambiguities in time-independent problems; (b) allows one to fix all the scales, not only the principal ones; (c) prepares ground for time-dependent problems.

According to the recipe (1.8), the original Schrödinger equation (1.1) becomes

$$-\frac{1}{2}\frac{\partial^2}{\partial x^2}\psi(x) + \frac{1}{2}x^2\psi(x) + \beta x^4\psi(x) = E\psi(x)\,. \tag{1.9}$$

In short, according to this recipe, instead of the original Eq. (1.1) we deal with its dimensionless form (1.9), ready for analytic or numerical work: the way to obtain (1.9) is to replace each of the parameters $\hbar^2/m$ and $m\omega^2$ by unity, and replace the parameter β by $\frac{\beta\hbar}{m^2\omega^3}$; the latter will further become a number, say 2.74, if some numerical answers are required:

$$\hbar^2/m \to 1$$
$$m\omega^2 \to 1$$
$$\beta \to \frac{\beta\hbar}{m^2\omega^3} \overset{\text{in numerics}}{\longrightarrow} 2.74 \rightsquigarrow \dots\,.$$

Imagine that we solved numerically the Eq. 1.9 for some value of β and obtained $E_{\text{g.s.}} = 1.14\dots$ To return to the usual system of units, we have to simply multiply this result by the energy scale $\hbar\omega$—i.e. the only parameter with units of energy that can be constructed out of $\hbar$, m, and ω:

$$\dots \rightsquigarrow E_{\text{g.s.}} = 1.14\dots \to E_{\text{g.s.}} = 1.14\dots \times \hbar\omega\,.$$

The true reason why the whole procedure looks mysterious at first is that formally speaking it severely abuses notations: for example, $E_{\text{g.s.}} = 1.14\dots$ and $E_{\text{g.s.}} = 1.14\dots \times \hbar\omega$ should be denoted by different symbols—but they are *not*. The practical advantages of this convention, however, compensate for the difficulties experienced at the learning stage.

1.1.2 *Dimensional analysis: the harmonic oscillator alone*

Now, let us try to produce a dimensional solution for the ground state energy of the harmonic oscillator alone:

$$-\frac{\hbar^2}{2m}\frac{\partial^2}{\partial x^2}\psi(x) + \frac{m\omega^2}{2}x^2\psi(x) = E\psi(x)\,.\tag{1.10}$$

<u>Solution:</u> The procedure goes as follows:

- *The principal units—the units of length and the units of energy:*

$$[\mathcal{L}]\,,\quad[\mathcal{E}]\,;$$

- *The input parameters and their units:*

$$[\eta] = [\mathcal{L}]^2[\mathcal{E}]$$
$$[\Upsilon] = [\mathcal{L}]^{-2}[\mathcal{E}]\,,$$

where again $\eta \equiv \hbar^2/m$, and $\Upsilon \equiv m\omega^2$;
- *The set of independent dimensionless parameters = $\emptyset$.* Indeed, assume there exists a dimensionless parameter P expressed as a product of powers of principal scales:

$$P = \text{const} \times \eta^{\xi_1}\Upsilon^{\xi_2}\,.$$

Its units are now

$$[P] = [\eta]^{\xi_1}[\Upsilon]^{\xi_2}$$
$$= \left([\mathcal{L}]^2[\mathcal{E}]\right)^{\xi_1}\left([\mathcal{L}]^{-2}[\mathcal{E}]\right)^{\xi_2}$$
$$= [\mathcal{L}]^{2\xi_1-2\xi_2}[\mathcal{E}]^{\xi_1+\xi_2}\,.$$

The analogue of Eq. (1.3) is

$$\hat{M}\begin{pmatrix}\xi_1\\\xi_2\end{pmatrix} = \begin{pmatrix}0\\0\end{pmatrix}\,,\tag{1.11}$$

where

$$\hat{M} = \begin{pmatrix}1 & 1\\2 & -2\end{pmatrix}\,.\tag{1.12}$$

Now, according to the rule (1.5), this problem has *no dimensionless parameters at all.* This is exactly the situation where dimensional analysis produces the most complete solutions, accurate up to an unknown numerical prefactor;

– *The principal scales—the length scale and the energy scale, examples of*:

$$\mathcal{L} = \sqrt{\frac{\hbar}{m\omega}}$$
$$\mathcal{E} = \hbar\omega \ .$$

They are exactly the same as in the full harmonic-quartic problem;
– *Solution for the unknown*:

$$[E_{\text{g.s.}}] = [\mathcal{E}] \Rightarrow E_{\text{g.s.}} = \text{const} \times \mathcal{E} = \text{const} \times \hbar\omega \ .$$

where const is a number of the order of unity. Its precise value is inaccessible for dimensional methods. Recall that the exact value of this constant is $1/2$.

Finally,

$$\boxed{E_{\text{g.s.}} \sim \hbar\omega}$$

1.1.3   *Order-of-magnitude estimate: full solution*

1.1.3.1   *Order-of-magnitude estimates vis-a-vis dimensional analysis*

The dimensional analysis exemplified above is a formal method. It requires almost no understanding of the physics of the problem at hand. It has the advantage of being potentially applicable when a poorly understood physical phenomenon is studied; for example, in Chapter 11, we will deploy the dimensional machinery to analyze properties of integrable partial differential equations, under an explicit assumption that the reader either did not have any prior experience with the Kortewg-de Vries and sine-Gordon equations, or tries his or her best to pretend he or she didn't. The drawback of dimensional analysis is its limited scope of applicability. It applies only to problems with a small enough number of not-of-the-order-unity input parameters.

Conversely, order-of-magnitude estimates require a deep understanding of the physics behind the problem. Two principal stages of the solution can be identified. At the first stage, the space of input parameters is split into regions of approximate applicability of simpler models. Often, the position of the boundary between the regimes becomes clear only after the simpler models are solved. At the second stage, these models are solved approximately: typically the geometry of the problem is rendered on a rectangular grid ("every closed trajectory is a rectangle, every angle is 90°, and integrals are height times width"), and all numbers of the order of unity are replaced by unity.

Our goal is to give an order of magnitude estimate of the ground state energy of the harmonic-quartic system (1.1).

A detailed <u>Solution</u> to this problem is presented on the next two pages.

1.1.3.2   *Harmonic vs. quartic regimes*

It is obvious that for sufficiently small values of the quartic potential strength β, the ground state energy will be dominated by the harmonic part. It is equally obvious that if the value of β is too large, the harmonic contribution to the potential can be neglected. What is less obvious is that the boundary $\beta_{\text{between domains of applicability}}$ between the domains of applicability ("regimes") of the harmonic and quartic theories must lie at a point $\beta_{\text{comparable predictions}}$ where these theories give comparable predictions:

$$\beta_{\text{between domains of applicability}} \sim \beta_{\text{comparable predictions}} \, .$$

To prove the latter assertion one should rely on the assumption that the ground state energy is a *continuous* function of the quartic strength β. Let us assume now that the boundary between regimes lies far below the point of comparable predictions, $\beta_{\text{between domains of applicability}} \ll \beta_{\text{comparable predictions}}$, and let us try to arrive at a contradiction. Indeed for $\beta \lesssim \beta_{\text{between domains of applicability}}$, the prediction of the harmonic theory applies, and for $\beta \gtrsim \beta_{\text{between domains of applicability}}$, the quartic theory applies. But since we are far from the point where the predictions of the two theories are comparable, at around $\beta_{\text{between domains of applicability}}$, the ground state energy as a function of β will have a discontinuity, that we disallowed. The identical reasoning applies for the case of $\beta_{\text{between domains of applicability}} \gg \beta_{\text{comparable predictions}}$. Thus, we arrive at a contradiction, *Q.E.D.*

The point of comparable predictions $\beta_{\text{comparable predictions}}$ is still unknown. But on both sides of this boundary, the models are much simpler than the original one: there, the ground state energy can be easily estimated. These estimates will be used to determine the boundary between regimes $\beta_{\text{between domains of applicability}}$.

1.1.3.3   *The harmonic oscillator alone*

To estimate the ground state energy of the harmonic oscillator,

$$\hat{H}_{\text{HO}} = \frac{\hat{p}^2}{2m} + \frac{m\omega^2}{2}x^2 \,, \tag{1.13}$$

assume that the ground state has a spatial width Δx. It is known that in the ground state, the Heisenberg uncertainty inequality becomes an approximate equality,

$$\Delta p \sim \frac{\hbar}{\Delta x} \,, \tag{1.14}$$

where Δp is the momentum width. The ground state energy can then be estimated as

$$E_{\text{g.s.}} \sim A\frac{\hbar^2}{m(\Delta x)^2} + Bm\omega^2(\Delta x)^2 \tag{1.15}$$

$$A \sim B \sim 1 \,, \tag{1.16}$$

where A and B are unknown entities of the order of unity. Now, it is also known that the ground state energy minimizes the expectation value of energy among all possible states. In particular, it should do so among states that yield an approximate Heisenberg equality (1.14). Thus the ground state energy is represented by the minimum of the energy (1.16) considered as a function of the width Δx. This minimum is given by

$$E_{\text{g.s., HO}} \sim \hbar\omega \,, \tag{1.17}$$

and it is reached at

$$(\Delta x)_{\text{g.s., HO}} \sim \sqrt{\frac{\hbar}{m\omega}} \,.$$

1.1.3.4   *The quartic oscillator alone*

The analysis of the ground state of a quartic oscillator,

$$\hat{H}_{\text{QO}} = \frac{\hat{p}^2}{2m} + \beta x^4 \,, \tag{1.18}$$

is identical to the analysis of the harmonic one. The result is

$$E_{\text{g.s., QO}} \sim \left(\frac{\hbar^2}{m}\right)^{2/3} \beta^{1/3} \,, \tag{1.19}$$

at

$$(\Delta x)_{\text{g.s., QO}} \sim \left(\frac{\hbar^2}{m}\right)^{1/6} \frac{1}{\beta^{1/6}} \,.$$

1.1.3.5   *The boundary between the regimes and the final result*

Setting the ground state energy predictions, (1.17) and (1.19), equal,

$$E_{\text{g.s., HO}} \sim E_{\text{g.s., QO}} ,$$

we get

$$\beta_{\text{between domains of applicability}} \sim \frac{m^2\omega^3}{\hbar} .$$

The final prediction for the ground state energy of a harmonic-quartic oscillator (1.1) reads

$$
E_{\text{g.s.}} \sim
\begin{cases}
\hbar\omega & \text{for } \beta \lesssim \dfrac{m^2\omega^3}{\hbar} \\[2ex]
\left(\dfrac{\hbar^2}{m}\right)^{2/3} \beta^{1/3} & \text{for } \beta \gtrsim \dfrac{m^2\omega^3}{\hbar}
\end{cases}
$$

1.1.4   *An afterthought: boundary between regimes from dimensional considerations*

It turns out that if we knew a priori there was a boundary between two regimes we would be able to get it on dimensional grounds. So the problem is: *find* $\beta_{\text{between domains of applicability}}$ *dimensionally.*

Solution: The dimensionless parameter P_1 (see (1.2)) is the only dimensionless number that controls how deep into the harmonic or quartic regime the system is. The transition between the regimes must happen at some special value of this parameter. A priori, for a non-negative dimensionless number, there exist only three special values: 0, 1, and $+\infty$. We know that the first corresponds to a deep harmonic regime, and the last corresponds to the quartic one. Thus, the transition between regimes must happen when the parameter P_1 reaches a value of the order of unity:

$$P_1\Big|_{\beta\sim\beta_{\text{between domains of applicability}}} \sim \frac{\hbar\beta_{\text{between domains of applicability}}}{m^2\omega^3} \sim 1 .$$

An estimate for the boundary between regimes immediately follows:

$$\beta_{\text{between domains of applicability}} \sim \frac{m^2\omega^3}{\hbar}$$

It is necessary to stress that this estimate does not imply that a harmonic-to-quartic transition exists. To show that, we have to go beyond dimensional analysis.

1.1.5   *A Gaussian variational solution*

Variational methods, introduced in Chapter 6, provide another simple but powerful tool that allows one to quickly estimate the result of a problem. Unlike dimensional and order-of-magnitude arguments, variational methods produce numerical answers, usually close to the exact result. *The assignment is: using a Gaussian variational ansatz, find the unknown function $\Phi(P_1)$ in Eq. 1.7.*

Solution The exact ground state $\psi_{\text{g.s., exact}}$ minimizes the energy functional

$$\mathcal{E}[\psi(\cdot)] = \int dx \left\{ \frac{\hbar^2}{2m}|\psi'(x)|^2 + \left(\frac{m\omega^2}{2}x^2 + \beta x^4 \right) |\psi(x)|^2 \right\}$$

on the space of smooth functions normalized to unity. The minimum of this functional on a one-parametric Gaussian manifold,

$$\psi(x,\,\sigma) \equiv \frac{1}{(2\pi\sigma^2)^{1/4}} e^{-x^2/(4\sigma^2)}$$

constitutes a Gaussian variational estimate for the true ground state energy.

The variational solution to the ground state energy is given by

$$E_{\text{g.s.}} \overset{\text{variational}}{=} \Phi(P_1) \times \hbar\omega \,,$$

with

$$\Phi(P_1) = \frac{6f(\phi(P_1))^2 + f(\phi(P_1))^3 + 1944P_1^2}{432f(\phi(P_1))P_1}$$

$$f(\phi) = \phi + \frac{1}{\phi} - 1$$

$$\phi(P_1) = (486P_1^2 + 18\sqrt{729P_1^4 - 3P_1^2} - 1)^{1/3}$$

$$P_1 = \frac{\hbar\beta}{m^2\omega^3}$$

This result, along with the purely harmonic and purely quartic predictions, is presented in Fig. 1.1. As expected, for $P_1 \gtrsim 2$, the variational value is fairly close to the quartic prediction. On the other hand, for $P_1 \lesssim .25$, the ground state energy is close to the one for the harmonic oscillator.

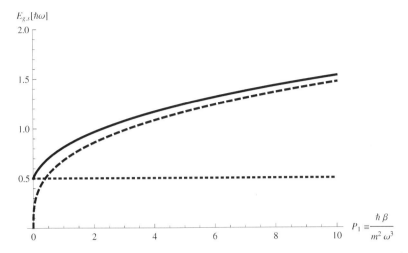

Fig. 1.1 A Gaussian variational solution for the ground state energy of a hybrid harmonic-quartic oscillator (1.1), as a function of the strength β of the quartic part of the potential (solid line). P_1 is the only dimensionless parameter of the problem. Exact energy for a purely harmonic oscillator (short-dashed line), and a Gaussian energy for a purely quartic oscillator (long-dashed line) are shown for comparisson.

Chapter 2

A dimensional estimate for the Planck temperature: a Case Study

2.1 Solved problems

2.1.1 *Estimating the Planck temperature*

No one knows for sure what quantum gravity is, or even whether gravity is quantum at all. Nonetheless, if it is quantum, dimensional analysis allows us to foresee at what temperature[1] quantum effects will emerge. In this estimate, we will assume that the properties of matter (electrons, etc.) are irrelevant, and the set of input parameters for the theory of quantum gravity is

gravitational constant G

Boltzmann constant, k_B

speed of light, c

Planck's constant, $\hbar$

<u>Solution:</u>

   – *The principal units—the units of length, time, mass, and temperature:*

$$[\mathcal{L}], \quad [\mathcal{T}], \quad [\mathcal{M}], \quad [\Theta]$$

$\#$ of units of measurement $\equiv N = 4 = (\#$ of rows of $\hat{q})$;

---

[1] ... an intensive quantity; for energy, one would need to specify, in addition, how many degrees of freedom are involved and what these degrees of freedom are.

– *The input parameters and their units*:

$$[G] = [\mathcal{L}]^3 / [\mathcal{M}] [\mathcal{T}]^2$$
$$[k_\mathrm{B}] = [\mathcal{M}] [\mathcal{L}]^2 / [\Theta] [\mathcal{T}]^2$$
$$[c] = [\mathcal{L}] / [\mathcal{T}]$$
$$[\hbar] = [\mathcal{M}] [\mathcal{L}]^2 / [\mathcal{T}]$$

# of input parameters $\equiv M = 4 = $ # of columns of $\hat{q}$;

Dimensional analysis matrix:

$$\hat{q} = \begin{array}{c|cccc} & G & k_\mathrm{B} & c & \hbar \\ \hline [\mathcal{L}] & +3 & +2 & +1 & +2 \\ [\mathcal{T}] & -2 & -2 & -1 & -1 \\ [\mathcal{M}] & -1 & +1 & 0 & +1 \\ [\Theta] & 0 & -1 & 0 & 0 \end{array}$$

– *The principal scales—length, time, mass, and temperature: set-up*:

$$\mathcal{L} = G^{\nu_{G,\mathcal{L}}} k_\mathrm{B}^{\nu_{k_\mathrm{B},\mathcal{L}}} c^{\nu_{c,\mathcal{L}}} \hbar^{\nu_{\hbar,\mathcal{L}}}$$
$$\mathcal{T} = G^{\nu_{G,\mathcal{T}}} k_\mathrm{B}^{\nu_{k_\mathrm{B},\mathcal{T}}} c^{\nu_{c,\mathcal{T}}} \hbar^{\nu_{\hbar,\mathcal{T}}}$$
$$\mathcal{M} = G^{\nu_{G,\mathcal{M}}} k_\mathrm{B}^{\nu_{k_\mathrm{B},\mathcal{M}}} c^{\nu_{c,\mathcal{M}}} \hbar^{\nu_{\hbar,\mathcal{M}}} \qquad (2.1)$$
$$\Theta = G^{\nu_{G,\Theta}} k_\mathrm{B}^{\nu_{k_\mathrm{B},\Theta}} c^{\nu_{c,\Theta}} \hbar^{\nu_{\hbar,\Theta}}$$

Our goal is to find the matrix

$$\hat{\nu} = \begin{pmatrix} \nu_{G,\mathcal{L}} & \nu_{G,\mathcal{T}} & \nu_{G,\mathcal{M}} & \nu_{G,\Theta} \\ \nu_{k_\mathrm{B},\mathcal{L}} & \nu_{k_\mathrm{B},\mathcal{T}} & \nu_{k_\mathrm{B},\mathcal{M}} & \nu_{k_\mathrm{B},\Theta} \\ \nu_{c,\mathcal{L}} & \nu_{c,\mathcal{T}} & \nu_{c,\mathcal{M}} & \nu_{c,\Theta} \\ \nu_{\hbar,\mathcal{L}} & \nu_{\hbar,\mathcal{T}} & \nu_{\hbar,\mathcal{M}} & \nu_{\hbar,\Theta} \end{pmatrix},$$

with

$$\text{\# of columns of } \hat{\nu} = N$$
$$\text{\# of rows of } \hat{\nu} = M.$$

For the equations (2.1) to make sense dimensionally, $\hat{\nu}$ must obey

$$\hat{q}\,\hat{\nu} = \hat{I}, \qquad (2.2)$$

where $\hat{I}$ is an N-by-N unit matrix.

– *Analysis*:

- First, recall that the rank of a matrix A, denoted rank A, is by definition its column rank, i.e., the dimension of its column space (which is the space spanned by the columns of A, when each column is considered an individual column vector). Equivalently, rank A is the total number of linearly independent columns. Further, a theorem guarantees that the row rank (the dimension of the row space) and the column rank must be equal.

- Here is the main result: Eq. (2.2) has at least one solution if and only if it has *full row rank* (all the row vectors form a linerly independent set). But the row rank equals the column rank, and $\hat{q}$ is an N-by-M matrix, so this is equivalent to

$$\text{rank}\,\hat{q} = N\,. \tag{2.3}$$

  The solution will be *unique* if and only if we additionally have $M = N$.

- We will prove this, but first, two side remarks: (i) if rank $\hat{q} = N$, then $N \leqslant M$, for the column rank of of a matrix cannot exceed the number of columns. (ii) An alternative way to say that rank $\hat{q} = N$ is this: $N \leqslant M$ and $\hat{q}$ has full rank.

- Now the proof. Consider first the equation $A\vec{x} = \vec{c}$, where A is an N-by-M matrix, and $\vec{x}$ and $\vec{c}$ are column vectors of dimensions M and N, respectively. Suppose A and $\vec{c}$ are given, and $\vec{x}$ is unknown. If we regard each column of A as a column vector, then, when we muliply A by a column vector, the result is a linear combination of A's column vectors. Thus, $A\vec{x} = \vec{c}$ will have a solution if and only if $\vec{c}$ is in the column space of A (i.e., it is representable as a linear combination of A's column vectors).

- Next, consider the matrix equation $AX = C$. It is equivalent to the set of equations $A\vec{x}_j = \vec{c}_j$, where $\vec{x}_j$ and $\vec{c}_j$ are the jth columns of X and C, repectively. Thus, $AX = C$ will have a solution if and only if each column of C is in the column space of A, meaning that the column space of C is contained in the column space of A. We will call this the *consistency condition*.

- Turning to our case of interest, $C = I_{N \times N}$, the N-by-N unit matrix, and $A = \hat{q}$, an N-by-M matrix. In this case, the consistency condition becomes that *the rank of A must be N*, and this is what we need to prove.

- Note that (i) the column space of $C = I_{N \times N}$ is $\mathbb{R}^N$, since the columns of $I_{N \times N}$ form a basis for it; and (ii) rank $A \leqslant N$, since the column rank equals row rank, and the row rank cannot exceed the number of rows. We will now show that the consistemcy condition holds if and only if rank $A = N$.

- Left-to-right: let's prove the contrapositive, i.e., that if rank $A \neq N$, then the consistency condition fails. Suppose rank $A \neq N$. Then either rank $A < N$ or rank $A > N$. The latter contradicts (ii), so we must have rank $A < N$. But then the column space of A has lower dimensionality than $\mathbb{R}^N$, so it cannot contain it, so the consistency condition fails.

- Right-to-left: suppose rank $A = N$. Note that each of the M columns of A is an N-dimensional column vector; in other words, each is a vector from $\mathbb{R}^N$. Since rank $A = N$, the dimension of the column space of A is N, so there are N columns of A which form a linearly independent set. But any set of N linearly indepenent vectors from $\mathbb{R}^N$ is a basis for it. Thus, these N columns span $\mathbb{R}^N$; thus, the column space of A contains $\mathbb{R}^N$, and we are done.

- Now we turn to uniqueness. Suppose (2.2) has a solution. Let's now show that it is *unique* if and only if $M = N$.

- First let's prove the following. Suppose $\hat{\nu}_1$ is a solution. $\hat{\nu}_2$ is also a solution (possibly the same, possibly different) if and only if each column of $\hat{\nu}_1 - \hat{\nu}_2$ is an element of the kernel of $\hat{q}$. (The kernel of $\hat{q}$, also called its *null space*, consists of all column vectors $\vec{v}$ such that $\hat{q}\,\vec{v} = 0$. It is easy to see that the kernel is a vector space). Left-to-right: if they are both solutions, then $\hat{q}\,\hat{\nu}_1 = \hat{I}$ and $\hat{q}\,\hat{\nu}_2 = \hat{I}$. Subtracting these two equations, we get $\hat{q}\,(\hat{\nu}_1 - \hat{\nu}_2) = 0_{N \times N}$ (an N-by-N matrix of zeroes). This means that each column of $\hat{\nu}_1 - \hat{\nu}_2$ is an element of the kernel of $\hat{q}$. Right-to-left: suppose $\hat{q}\,(\hat{\nu}_1 - \hat{\nu}_2) = 0_{N \times N}$. Then $\hat{q}\,\hat{\nu}_1 = \hat{q}\,\hat{\nu}_2$. But $\hat{\nu}_1$ is a solution, so $\hat{q}\,\hat{\nu}_1 = \hat{I}_{N \times N}$. So $\hat{q}\,\hat{\nu}_2 = \hat{I}_{N \times N}$ as well, and we are done.

- We next prove that, provided (2.2) has a solution, it will be unique if and only if the *nullity* of $\hat{q}$ (the dimension of the kernel of $\hat{q}$) is 0, i.e., the kernel consists only of the zero vector.

- Left-to-right: we'll prove the contrapositive, i.e., that if the nullity of $\hat{q}$ is 1 or more, then the solution is not unique. Let B be a matrix (of the same dimensions as $\hat{\nu}$) all of whose columns are elements of the kernel of $\hat{q}$. Since the kernel of $\hat{q}$ has a dimension of 1 or more, there is at least one non-zero vector $\vec{w}$ such that $\hat{q}\,\vec{w} = 0$. Assume that at least one of the columns of B is this vector. Then $B \neq 0_{N \times N}$ and $\hat{q}\,B = 0_{N \times N}$. We assumed (2.2) has at least one solution; let $\hat{\nu}$ that solution. Then both $\hat{\nu}$ and $\hat{\nu} + B$ are solutions (since $\hat{q}\,(\hat{\nu} + B) = \hat{q}\,\hat{\nu} + \hat{q}\,B - 0_{N \times N} \mid 0_{N \times N} - 0_{N \times N}$), but, since $B \neq 0_{N \times N}$ they are not equal. So the solution isn't unique.

- Right-to-left: suppose the kernel of $\hat{q}$ has dimension 0. We have to prove that if $\hat{\nu}_1$ and $\hat{\nu}_2$ are both solutions, then $\hat{\nu}_1 = \hat{\nu}_2$. So, suppose $\hat{\nu}_1$ and $\hat{\nu}_2$ are both solutions. Then, as we have seen above, each column of $\hat{\nu}_1 - \hat{\nu}_2$ must be an element of the kernel of $\hat{q}$. But the kernel of $\hat{q}$ has dimension 0; so it consists of just the zero vector; so $\hat{\nu}_1 - \hat{\nu}_2 = 0_{N \times N}$; so $\hat{\nu}_1 = \hat{\nu}_2$, so the solution is unique.

- The final step is to prove that the nullity of $\hat{q}$ is 0 if and only if $N = M$. We will use the rank-nullity theorem: the sum of the rank and the nullity equals the number of columns. Thus, the nullity is zero if and only if the rank equals the number of columns. In our case, the rank of $\hat{q}$ is N (since we are assuming there is at least one solution), while the number of columns is M. Thus, the nullity of $\hat{q}$ is 0 if and only if $N = M$, and we are done.

- The condition that the solution exists and is unique is thus that $\hat{q}$ is an N-by-N matrix of rank N, which means that

$$\det \hat{q} \neq 0\,.$$

In that case, we have

$$\hat{\nu} = \hat{q}^{-1}\,. \tag{2.4}$$

In our particular example, $\hat{q}$ is a 4-by-4 matrix, and its determinant is indeed non-zero: $\det \hat{q} = 2$.

If the condition (2.3) is violated, it means that some of the principal scales do not exist. This, in turn, usually leads to nontrivial predictions. For example, imagine a world where an upper bound on speed, the speed of light c, is absent. In this case, the condition (2.3) is violated. In particular, no observable with dimensions of temperature can be constructed there. Looking ahead, the Planck temperature T_P of our world (see below) would drift to infinity

$$(T_P)_{new} \to \infty\,;$$

in a world with no special relativity, gravity is never quantized.
- If (2.3) is satisfied, the fundamental scales (2.1) do exist. If

$$\operatorname{rank}\hat{q} = N = M\,,$$

then these scales are unique. Otherwise, the scales are fixed up to a multiplication by an arbitrary function of $M - \operatorname{rank}\hat{q}$ dimensionless combinations of the input parameters. For example, imagine that in the future, a new fundamental law of Nature is discovered, along with a new fundamental constant c_2, of a dimensionality of speed; there $M - \operatorname{rank}\hat{q} = 1$. In that case, the Planck temperature T_P of our world (again, see below) would need to be replaced by a *much less informative estimate,*

$$(T_P)_{new} \sim T_P f(c_2/c)\,,$$

where $f(\xi)$ is an arbitrary function.
- *The principal scales—length, time, mass, and temperature: result*: Using (2.4), we get

$$\hat{\nu} = \begin{array}{c|cccc} & [\mathcal{L}] & [\mathcal{T}] & [\mathcal{M}] & [\Theta] \\ \hline G & +\frac{1}{2} & +\frac{1}{2} & -\frac{1}{2} & -\frac{1}{2} \\ k_B & 0 & 0 & 0 & -1 \\ c & -\frac{3}{2} & -\frac{5}{2} & +\frac{1}{2} & +\frac{5}{2} \\ \hbar & +\frac{1}{2} & +\frac{1}{2} & +\frac{1}{2} & +\frac{1}{2} \end{array}$$

or

$$\mathcal{L} = \sqrt{\frac{\hbar G}{c^3}}$$

$$\mathcal{T} = \sqrt{\frac{\hbar G}{c^5}}$$

$$\mathcal{M} = \sqrt{\frac{\hbar c}{G}}$$

$$\Theta = \frac{1}{k_B}\sqrt{\frac{\hbar c^5}{G}}\,;$$

– Solution for the unknowns:

$$[T_{\mathrm{P}}] = [\Theta] \Rightarrow T_{\mathrm{P}} \sim \Theta\,.$$

We get, finally,

$$T_{\mathrm{P}} \sim \frac{1}{k_{\mathrm{B}}}\sqrt{\frac{\hbar c^5}{G}}$$

Chapter 3

Bohr-Sommerfeld Quantization

3.1 Solved problems

3.1.1 *Ground state energy of a harmonic oscillator*

Consider the Schrödinger equation for a particle in a harmonic oscillator:

$$-\frac{\hbar^2}{2m}\frac{\partial^2}{\partial x^2}\psi(x) + \frac{m\omega^2}{2}x^2\psi(x) = E\psi(x)\,.$$

(a) *Using the Heisenberg uncertainty principle and minimization of energy, estimate the ground state energy.*

Solution: The ground state minimizes the Heisenberg uncertainty:

$$\Delta p\,\Delta x \sim \hbar\,.$$

Hence, as a function of the momentum uncertainty δp, the ground state energy becomes

$$E_{\text{g.s.}} \sim A\frac{\Delta p^2}{m} + B\frac{\hbar^2 m\omega^2}{\Delta p^2}\,,$$

where A and B are dimensionless constants of the order of unity. Minimizing $E_{\text{g.s.}}$ with respect to Δp, one gets

$$\Delta p \sim \sqrt{\hbar m\omega}\,,$$

and

$$\boxed{E_{\text{g.s.}} \sim \hbar\omega}$$

(b) *The same as the above, but using dimensional analysis.*

Solution: The input parameters are $\hbar^2/m \equiv \eta$ and $m\omega^2 \equiv \Upsilon$ are given by $[\eta] = \mathcal{E}\mathcal{L}^2$ and $[\Upsilon] = \mathcal{E}/\mathcal{L}^2$. The units for the ground state energy $E_{\text{g.s.}}$ are $[E_{\text{g.s.}}] = \mathcal{E}$. Thus,

$$E_{\text{g.s.}} \sim \sqrt{\eta \Upsilon} \sim \hbar\omega$$

3.1.2 Spectrum of a harmonic oscillator

(a) *Using the Bohr-Sommerfeld quantization rule,*

$$\oint p(x|E_n)\,dx = 2\pi\hbar(n + 1/2) \qquad (3.1)$$

$$n = 0,\ 1,\ 2,\ \dots\ ,$$

where

$$p(x|E) \equiv \pm\sqrt{2m(E - V(x))}$$

$$\oint p(x|E)\,dx \equiv 2\int_{x_{min}}^{x_{max}} |p(x|E)|\,dx\ ,$$

determine the energy spectrum of a harmonic oscillator;

Solution: The WKB integral

$$A = \oint \sqrt{2m\left(E - \frac{m\omega^2 x^2}{2}\right)}\,dx$$

is nothing else but the area, $A = \pi r_x r_p$, of an ellipse with the half-axes $r_x = \sqrt{2E/(m\omega^2)}$ and $r_p = \sqrt{2mE}$. Thus, this integral is given by $A = 2\pi E/\omega$. Finally

$$E_n = \hbar\omega\left(n + \frac{1}{2}\right)$$

(b) *Pretending you do not know how to compute the integral in Eq. 3.1, estimate E_n.*

Solution: Replacing the ellipse by a rectangle, we get $A \approx r_x r_p \sim \sqrt{E/(m\omega^2)}\sqrt{mE} = E/\omega$. Finally, we get

$$E_n \sim \hbar\omega(n + 1/2)$$

(c) *Pretending you do not even know what an "integral" is—just that its symbol "$\oint$" is a dimensionless entity—estimate E_n using dimensional analysis.*

Solution: The Bohr-Sommerfeld quantization rule again reads

$$\oint \sqrt{2m \left(E_n - \frac{m\omega^2 x^2}{2} \right)} \, dx = 2\pi\hbar(n + 1/2)$$

$$n = 0,\, 1,\, 2,\, \ldots\,;\tag{3.2}$$

but this time around, we pretend that the integral is some obscure, poorly understood mathematical operation, and all we know is units the participants of the Eq. (3.2) are measured in.

Notice however that the state index n enters (3.2) *only* as one of the two factors of the product $\hbar(n + 1/2)$. This allows us to introduce a new dimensionful parameter, $\tilde{\hbar}$. Furthermore, we realize that only two independent dimensionful parameters are needed: $\tilde{\eta} \equiv \tilde{\hbar}^2/m$ and $\Upsilon \equiv m\omega^2$. The quantization rule now reads:

$$\oint \sqrt{2 \left(E_n - \frac{\Upsilon x^2}{2} \right)} \, dx = 2\pi\sqrt{\tilde{\eta}}$$

$$n = 0,\, 1,\, 2,\, \ldots\quad.$$

Performing dimensional analysis we get

- *The principal units—the units of length and the units of energy*:

$$[\mathcal{L}]\,,\quad [\mathcal{E}]\,;$$

- *The input parameters and their units*:

$$[\tilde{\eta}] = [\mathcal{L}]^2\,[\mathcal{E}]$$
$$[\Upsilon] = [\mathcal{L}]^{-2}\,[\mathcal{E}]\,;$$

- *The set of independent dimensionless parameters* $= \emptyset$;
- *The principal scales—the length scale and the energy scale, examples of*:

$$\mathcal{L} = \sqrt{\frac{\hbar}{m\omega}}$$
$$\mathcal{E} = \hbar\omega\,;$$

- *Solution for the unknown*:

$$E_n \sim \sqrt{\tilde{\eta}\Upsilon} = \tilde{\hbar}\omega\,.$$

Finally,

$$E_n \sim \hbar\omega(n + 1/2)$$

3.1.3 WKB treatment of a "straightened" harmonic oscillator

Consider a "straightened" harmonic oscillator potential

$$V(x) = \begin{cases} \frac{m\omega^2(x+L/2)^2}{2} & \text{for } x < -L/2 \\ 0 & \text{for } -L/2 \leq x < +L/2 \\ \frac{m\omega^2(x-L/2)^2}{2} & \text{for } x \geq +L/2 \end{cases},$$

see Fig. 3.1.

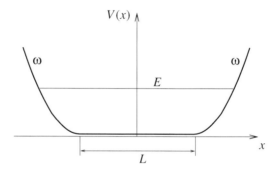

Fig. 3.1   Straightened harmonic oscillator.

Derive a WKB expression for the energy levels E_n.
Useful integral:

$$\int_0^1 \sqrt{1 - t^2}\, dt = \frac{\pi}{4}.$$

Solution: The Bohr-Sommerfeld quantization rule reads:

$$S(E_n) = 2\pi\hbar\left(n + \frac{1}{2}\right) \tag{3.3}$$

$$n = 0, 1, 2, \ldots,$$

where

$$S(E) = 2\int_{x_1}^{x_2} dx\, \sqrt{2m(E - V(x))}$$

is the so-called classical full-cycle abbreviated action. Here x_1 and x_2 are the left and right classical turning points respectively.

In our case

$$S(E) = S_{\text{HO}}(E) + S_{\text{free}}(E),$$

where

$$S_{\text{HO}}(E) = 2 \int_{-L/2-\sqrt{2E/(m\omega^2)}}^{-L/2} dx \sqrt{2m(E - m\omega^2(x+L/2)^2/2)}$$

$$+2 \int_{+L/2}^{+L/2+\sqrt{2E/m\omega^2}} dx \sqrt{2m(E - m\omega^2(x-L/2)^2/2)}$$

$$= 2 \int_{-\sqrt{2E/(m\omega^2)}}^{+\sqrt{2E/(m\omega^2)}} dx \sqrt{2m(E - m\omega^2 x^2/2)}$$

$$= \frac{2\pi E}{\omega}$$

and

$$S_{\text{free}}(E) = 2\sqrt{2mE}L.$$

Substitution of the above to the quantization rule (3.3) leads to

$$E_n = E_{\text{HO},n} + 2\left(\mathcal{E} - \sqrt{\mathcal{E}(E_{\text{HO},n} + \mathcal{E})}\right)$$

$$n = 0, 1, 2, \ldots,$$

where

$$E_{\text{HO},n} = \hbar\omega\left(n + \frac{1}{2}\right)$$

$$\mathcal{E} = \frac{1}{2\pi^2} m\omega^2 L^2$$

Interestingly, in the limit of the infinitely strong oscillator where $\omega \to \infty$, the energy spectrum reduces to

$$E_n \overset{\omega\to\infty}{\approx} \frac{\hbar^2}{2m}\left(\frac{\pi}{L}\right)^2 \left(n + \frac{1}{2}\right)^2$$

$$n = 0, 1, 2, \ldots,$$

which is nothing else but the spectrum of a box with "soft walls"—sharp boundaries where, at the same time, the spatial derivative of potential remains finite. Compare with the hard-wall box result,

$$E_{\text{hard-wall box}, n} = \frac{\hbar^2}{2m} \left(\frac{\pi}{L}\right)^2 (n+1)^2$$
$$n = 0, 1, 2, \dots .$$

Note also, that when we remove the flat region of our potential (i.e. set $L = 0$) the spectrum reduces to the usual spectrum of the harmonic oscillator:

$$E_n \overset{L=0}{=} \hbar\omega \left(n + \frac{1}{2}\right)$$
$$n = 0, 1, 2, \dots ,$$

3.1.4 *Ground state energy of power-law potentials*

Consider the Schrödinger equation for a particle in a "$2q$-tic" potential:

$$-\frac{\hbar^2}{2m} \frac{\partial^2}{\partial x^2} \psi(x) + K_q x^{2q} \psi(x) = E\psi(x)$$
$$q = 1, 2, 3, \dots .$$

(a) *Using the Heisenberg uncertainty principle and minimization of energy, estimate the ground state energy.*

Solution: The ground state minimizes the Heisenberg uncertainty:

$$\Delta p \, \Delta x \sim \hbar .$$

Hence, as a function of the momentum uncertainty δp, the ground state energy becomes

$$E_{\text{g.s.}} \sim A \frac{\Delta p^2}{m} + B \frac{\hbar^{2q} K_q}{\Delta p^{2q}} ,$$

where A and B are dimensionless constants of the order of unity. Minimizing $E_{\text{g.s.}}$ with respect to Δp, one gets

$$\Delta p \sim \hbar^{\frac{q}{q+1}} m^{\frac{1}{2(q+1)}} K_q^{\frac{1}{2(q+1)}} ,$$

and

$$\boxed{E_{\text{g.s.}} \sim \left(\frac{\hbar^2}{m}\right)^{\frac{q}{q+1}} K_q^{\frac{1}{q+1}}}$$

(b) *The same as the above, but using dimensional analysis.*

Solution: The input parameters are $\hbar^2/m \equiv \eta$ and $K_q \equiv \Upsilon$. Their units are given by $[\eta] = \mathcal{E}\mathcal{L}^2$ and $[\Upsilon] = \mathcal{E}/\mathcal{L}^{2q}$. The units for the ground state energy $E_{\text{g.s.}}$ are $[E_{\text{g.s.}}] = \mathcal{E}$. Thus,

$$E_{\text{g.s.}} \sim \eta^{\frac{q}{q+1}} \Upsilon^{\frac{1}{q+1}} \sim \left(\frac{\hbar^2}{m}\right)^{\frac{q}{q+1}} K_q^{\frac{1}{q+1}}$$

3.1.5 *Spectrum of power-law potentials*

Consider the Schrödinger equation for a particle in a "$2q$-tic" potential:

$$-\frac{\hbar^2}{2m}\frac{\partial^2}{\partial x^2}\psi(x) + K_q x^{2q}\psi(x) = E\psi(x).$$

(a) *Give an order-of-magnitude estimate for the spectrum E_n; use the WKB quantization rule as a starting point.*

Solution: WKB quantization reads

$$\oint p(x|E_n)\,dx = 2\pi\hbar\left(n + \frac{1}{2}\right).$$

Estimating the WKB integral as

$$\oint p(x|E_n)\,dx \sim \underbrace{\sqrt{mE_n}}_{\Delta p}\underbrace{(E_n/K_q)^{\frac{1}{2q}}}_{\Delta x},$$

we get

$$\sqrt{m}E_n^{\frac{q+1}{2q}}\frac{1}{K_q^{\frac{1}{2q}}} \sim \hbar(n + 1/2),$$

or

$$E_n \sim \left(\frac{\hbar^2}{m}\right)^{\frac{q}{q+1}} K_q^{\frac{1}{q+1}}(n + 1/2)^{\frac{2q}{q+1}}$$

(b) *Estimate the spectrum E_n using dimensional analysis applied to the WKB quantization rule.*

Solution: In this case, the method of solution is completely analogous to the one used in Problem 3.1.2. Consider WKB quantization at large n:

$$\oint \sqrt{2m(E_n - K_q x^{2q})}\, dx = 2\pi\hbar(n + 1/2)\,.$$

We can identify two input parameters: $(\hbar(n + 1/2))^2/m \equiv \tilde{\eta}$ and K_q, with $[\tilde{\eta}] = \mathcal{E}\mathcal{L}^2$ and $[K_q] = \mathcal{E}/\mathcal{L}^{2q}$. Here, K_q plays a role of Υ for the harmonic oscillator (Problem 3.1.2). The eigenenergies E_n, which have the dimension of energy, are given by

$$\boxed{E_n \sim \left(\frac{\hbar^2}{m}\right)^{\frac{q}{q+1}} K_q^{\frac{1}{q+1}} (n + 1/2)^{\frac{2q}{q+1}} \quad (3.4)}$$

Remark: A rigorous WKB quantization gives

$$E_n = C_q \left(\frac{\hbar^2}{m}\right)^{\frac{q}{q+1}} K_q^{\frac{1}{q+1}} (n + 1/2)^{\frac{2q}{q+1}}\,,$$

where

$$C_q = \left(\frac{\pi}{2}\right)^{\frac{q}{q+1}} \left(\frac{\Gamma[\frac{3q+1}{2q}]}{\Gamma[\frac{2q+1}{2q}]}\right)^{\frac{2q}{q+1}}, \quad (3.5)$$

and $\Gamma[z]$ is the gamma-function.

3.1.6   *The number of bound states of a diatomic molecule.*

Consider two atoms with masses m_1 and m_2, interacting via a "hard-core Van der Waals potential",

$$V(r) = \begin{cases} -\dfrac{C_6}{r^6}\,, & r \geq R \\ +\infty\,, & r < R \end{cases}. \quad (3.6)$$

(a) *Using dimensional analysis and the WKB approximation, estimate the total number N of the s-wave bound states ($l = 0$) of a diatomic molecule consisting of these two atoms.*

Recall that the relative motion is the one of a single particle of the so-called reduced mass $\mu = m_1 m_2/(m_1 + m_2)$, moving in the potential (3.6).

Solution: The only dimensionless combination one can form out of the set $\hbar^2/\mu$, C_6, and R is

$$\xi \equiv \frac{\hbar^2 R^4}{\mu C_6}\,.$$

Therefore the (dimensionless) number N can only have the form

$$N = \Phi\left(\frac{\hbar^2 R^4}{\mu C_6}\right), \tag{3.7}$$

where $\Phi(\xi)$ is an unknown dimensionless function.

At the level of the Schrödinger equation, Eq. 3.7 constitutes the absolute maximum of the information one can extract from the dimensional analysis. However, at the WKB level, the level indices enter only as a factor in the product $\hbar(n - 1/4)$ (where $n = 1, 2, 3, \ldots$),[1] the physical reason being that in the limit $n \gg 1$, all the appearances of Planck's constant must be absorbed in the classical action, $I = \oint p\,dx = \hbar(n - 1/4)$. On the other hand, N is an upper bound for such indices, and therefore, any relation it is involved in should only contain functions of $\hbar(N - 1/4)$ as a whole. This limits $\Phi(\xi)$ to

$$\Phi(\xi) = \text{const} \times \frac{1}{\sqrt{\xi}} + 1/4.$$

Thus, finally

$$\boxed{N - 1/4 \sim \frac{\sqrt{mC_6}}{\hbar R^2}}$$

(b) *Compute this number exactly, within WKB.*

Recall that for the hard-wall/soft-wall combination the Bohr-Sommerfeld rule reads

$$\oint p(x \mid E_n)\,dx = 2\pi\hbar\left(n + \frac{3}{4}\right), \qquad n = 0, 1, 2, \ldots. \tag{3.8}$$

Solution: The upper bound n_{max} can be obtained from the Bohr-Sommerfeld quantization rule (3.8) by setting E_n to zero. The integral can then be easily computed, and it gives

$$\boxed{N = \left\lfloor \frac{1}{\sqrt{2\pi}}\frac{\sqrt{mC_6}}{\hbar R^2} + \frac{1}{4} \right\rfloor}$$

where $\lfloor\ldots\rfloor$ is the floor function.

---

[1] Here, we apply the quantization rule (3.20) corresponding to one "hard" and one "soft" turning point.

3.1.7   *Coulomb problem at zero angular momentum*

The zero-angular-momentum motion of an electron in the field of an atomic nucleus is described by the one-dimensional Schrödinger equation

$$-\frac{\hbar^2}{2m}\frac{\partial^2}{\partial r^2}\chi(r) - \frac{\alpha}{r}\chi(r) = E\chi(r)$$

$$0 < x < \infty \qquad\qquad (3.9)$$

$$\chi(r \to 0) = 0 \,,$$

where $\alpha = Ze^2$, e is the electron charge, and Z is the number of protons in the nucleus.[2]

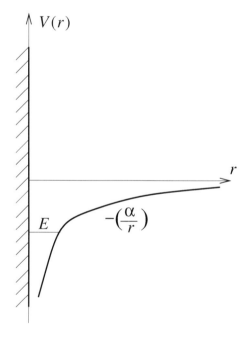

Fig. 3.2   Coulomb problem at zero angular momentum.

---

[2]The effective one-dimensional wavefunction $\chi(r)$ is related to actual solutions $\Psi(r, \Theta, \phi)$ of the three-dimensional Coulomb problem $-\frac{\hbar^2}{2m}\delta\Psi(r, \Theta, \phi) - \frac{\alpha}{r}\Psi(r, \Theta, \phi) = E\Psi(r, \Theta, \phi)$ as $\Psi(r, \Theta, \phi) = r^{-1}\chi_l(r)Y_{l,\,m}(\Theta, \phi)$, where l and m are the azimuthal and magnetic quantum numbers respectively, and $Y_{l,\,m}$ are the spherical harmonics.

(a) *Find the ground state energy E_1 in three ways:*

(i) *From the dimensional analysis;*

Solution:

- *The principal units—the units of length and the units of energy:*

$$[\mathcal{L}], \quad [\mathcal{E}];$$

- *The input parameters and their units:*

$$\left[\tilde{\eta} \equiv \hbar^2/m\right] = [\mathcal{L}]^2 [\mathcal{E}]$$
$$[\alpha] = [\mathcal{L}][\mathcal{E}];$$

- *The set of independent dimensionless parameters $= \emptyset$;*
- *The principal scales—the length scale and the energy scale, examples of:*

$$\mathcal{L} = \frac{\hbar}{m\alpha}$$
$$\mathcal{E} = \frac{m\alpha^2}{\hbar^2};$$

- *Solution for the unknown:*

$$E_1 \sim \mathcal{E} = \frac{m\alpha^2}{\hbar^2}.$$

We get finally

$$\boxed{E_1 \sim \frac{m\alpha^2}{\hbar^2}}$$

(ii) *From the Heisenberg uncertainty principle:*

(α) *Estimate the spatial extent Δr of the motion at a given energy E;*

(β) *Estimate the momentum spread Δp using dimensional analysis;[3]*

(γ) *Obtain the ground state energy E_1 using the fact that the Heisenberg inequality $\Delta p \, \Delta r \gtrsim \hbar$ becomes an (approximate) equality $\Delta p \, \Delta r \sim \hbar$ in the ground state.*

---

[3]Estimating Δp from the phase-space trajectory (as we do usually) is not straightforward, because the momentum diverges at $r = 0$.

Solution:

(α) A classical estimate for the apogee gives

$$\frac{\alpha}{\Delta r} \sim E \quad \Rightarrow \quad \Delta r \sim \frac{\alpha}{E}.$$

(β) A classical dimensional analysis (no $\hbar$, energy E instead) gives

$$\Delta p \sim \sqrt{mE}.$$

(γ) In the ground state, the Heisenberg inequality becomes close to an equality:

$$\Delta p(E_1)\,\Delta r(E_1) \sim \hbar.$$

Thus

$$\boxed{E_1 \sim \frac{m\alpha^2}{\hbar^2}}$$

(iii) *From the Bohr-Sommerfeld rule:*[4]

$$\oint p(r)\,dr = 2\pi\hbar(n+\delta) \tag{3.10}$$

$$n = 0,\ 1,\ 2,\ 3,\ \ldots\,.$$

Useful integral:

$$\int_0^1 \sqrt{1/t - 1}\ dt = \frac{\pi}{2}.$$

---

[4]We deliberately leave δ undefined. The question of a proper WKB quantization of a three-dimensional radially symmetric potential, including Coulomb problems at both non-zero and zero angular momenta, has a long and convoluted history. A rigorous analysis of the question is presented in a comprehensive book on WKB by Heading [M. A. Heading, *An Introduction to Phase-Integral Methods* (Wiley, New York, 1962)]. The currently most accepted quantization recipe is based on the classic paper by Langer [R. E. Langer, On the connection formulas and the solutions of the wave equation, *Physical Review* **51**, 669 (1937)]. In short, Langer's WKB prescription for a potential $V(r)$ at angular momentum l is as follows:

$$\oint p_{\text{eff.}}(r)\,dr = 2\pi\hbar(n_r + 1/2)$$

$$n_r = 0,\ 1,\ 2,\ 3,\ \ldots$$

$$p_{\text{eff.}}(r) \equiv \sqrt{2m(E_{n_r} - V_{\text{eff.}}(r))}$$

$$V_{\text{eff.}}(r) \equiv V(r) + \frac{\hbar^2(l+1/2)^2}{2mr^2}\,;$$

notice the replacement $l(l+1) \to (l+1/2)^2$ (the so-called Langer correction). The rule applies to both regular and singular potentials for any value of the angular momentum, including $l = 0$.

Solution:

$$\int_0^{r_2(E)} dr \sqrt{2m(-|E| + \alpha/r)} = \sqrt{2m|E|}\, r_2(E) \int_0^1 dt \sqrt{1/t - 1}$$

$$= \frac{\sqrt{2m}\alpha}{\sqrt{|E|}} \frac{\pi}{2}.$$

Here, $r_2(E) = \alpha/|E|$ is the right turning point (the apogee point for the Coulomb case). On the other hand

$$\int_0^{r_2(E_n)} \sqrt{2m(-|E_n| + \alpha/r)} = \pi\hbar(n + \delta).$$

We get

$$E_n = -\frac{(m\alpha^2/\hbar^2)}{2(n + \delta)^2}$$

$$n = 0, 1, 2, 3, \dots .$$

For the ground state energy we have

$$\boxed{E_1 = \left(-\frac{1}{2}\right) \frac{m\alpha^2}{\hbar^2} \frac{1}{\delta^2}}$$

(b) *Verify that*

$$\chi_1(r) = \frac{2}{\sqrt{a_Z}} (r/a_Z) \exp(-r/a_Z) \tag{3.11}$$

obeys the Schrödinger equation (3.9). Here, $a_Z = a_B/Z$, and $a_B = \hbar^2/(me^2)$ is the Bohr radius. Find the corresponding energy E_1. Since (3.11) has the least number of the nodes allowed by the boundary conditions in (3.9), it corresponds to the ground state, and thus the energy E_1 you will infer from (3.11) is the exact ground state energy. Compare the exact E_1 you will get with the WKB prediction from sub-problem (a)-(iii).

Solution: Substitution of the state (3.11) into the Schrödinger equation (3.9) makes the latter an equality for

$$\boxed{E_1 - \left(-\frac{1}{2}\right) \frac{m\alpha^2}{\hbar^2}}$$

3.1.8   *Quantization of angular momentum from WKB*

Consider a particle of mass m moving on a surface of a sphere of radius R. The energy of the particle is determined by its azimuthal quantum number l:

$$E_{l,m} = \frac{\hbar^2 l(l+1)}{2mR^2}\,.$$

At large energies (thus at short de-Broglie wavelengths), one may assume that the particle's wavefunction becomes less sensitive to the curvature of the sphere, and the problem can be approximately replaced by a *flat* billiards of the same area as the sphere. According to this interpretation,

$$P \equiv \frac{\hbar l}{R}$$

becomes the magnitude of the linear momentum of the particle, and the energy,

$$E_{l,m} \overset{E_{l,m} \gg \hbar^2/(mR^2)}{\approx} \frac{(\hbar l/R)^2}{2m} = \frac{P^2}{2m}\,,$$

becomes reinterpreted as the particle's kinetic energy. **(a)** *Using the multi-dimensional WKB quantization rule (see Weyl law in Sec. 3.3.2), determine $\bar{N}(E)$—the number of eigenstates with energy below or equal to some energy E;*

<u>Solution:</u> For a given energy E, the phase space volume occupied by points with energy less than or equal to E is

$$W(\text{energy} \leq E) = \underbrace{\text{2D-Volume(surface of 3D-sphere, coooordinates)}}_{4\pi R^2}$$
$$\times \underbrace{\text{2D-Volume(2D-ball, momenta)}}_{\pi P^2}$$
$$= 4\pi^2 R^2 P^2\,.$$

Now, according to the Weyl law,

$$\bar{N}(E) \overset{E \gg \hbar^2/(mR^2)}{\approx} \frac{W(\text{energy} \leq E)}{(2\pi\hbar)^2}\,.$$

We get:

$$\boxed{\begin{array}{l} \bar{N}(E) \overset{E \gg \hbar^2/(mR^2)}{\approx} 2\,\dfrac{E}{\hbar^2/(mR^2)} \\[2ex] \phantom{\bar{N}(E)} \overset{E \gg \hbar^2/(mR^2)}{\approx} l^2 \end{array}}$$

where $E = (\hbar l/R)^2/2m \overset{E \gg \hbar^2/(mR^2)}{\approx} E_{l,m}$

(b) *Using the exact quantization of the motion on a sphere, $E_{l,m} = \hbar^2 l(l+1)/(2mR^2)$, $l = 0, 1, \ldots$; $m = -l, -l+1, \ldots, +l$, find an expression for $\bar{N}_l$—the number of angular momentum eigenstates with the azimuthal quantum number l' less than or equal to l and compare it with your result from part (a).*

Solution:

$$\bar{N}_l = \sum_{l'=0}^{l} \sum_{m=-l}^{+l} 1 \, ,$$

or

$$\boxed{\bar{N}_l = (l+1)^2}$$

The agreement between (a) and (b) is quite remarkable.

3.1.9 *From WKB quantization of 4D angular momentum to quantization of the Coulomb problem*

It is not well-known[5] that there exists a map between the eigenstates $|n, l, m\rangle$ of the (three-dimensional) Coulomb problem $\hat{H} = p^2/(2m) - \alpha/r$ and the eigenstates $|l_{4D}, l_{3D}, m\rangle$ of a particle moving on a surface of a *four-dimensional* sphere of radius R. In this map, the quantum number n is related to the four-dimensional azimuthal quantum number l_{4D} as $n = l_{4D} + 1$. Likewise, $l = l_{3D}$. The energy on the 4D sphere is quantized as

$$E_{l_{4D}, l_{3D}, m} = \frac{\hbar^2 l_{4D}(l_{4D} + 2)}{2mR^2}. \tag{3.12}$$

Similarly to the three-dimensional case of Problem 3.1.8, we would assume that high energies, the 3D "surface" of the 4D-sphere will look almost flat, and the surface can be un-bent to a 3D Cartesian billiards. Again, analogously to the Problem 3.1.8 case,

$$P \equiv \frac{\hbar l_{4D}}{R}$$

---

[5]V. Fock, Zur Theorie des Wasserstoffatoms, *Zeitschrift für Physik* **98**, 145 (1935) [English translation: V. Fock, On the theory of the hydrogen atom, in A. Bohm, Y. Ne'eman, and A. O. Barut (eds.), *Dynamical Groups and Spectrum Generating Algebras*, vol. 1 (World Scientific, Singapore, 1988), p. 411.]

should be interpreted as momentum, and

$$E_{l_{4D},l_{3D},m} \overset{E_{l_{4D},l_{3D},m} \gg \hbar^2/(mR^2)}{\approx} \frac{(\hbar l_{4D}/R)^2}{2m} = \frac{P^2}{2m}$$

as kinetic energy.

(a) *Using Weyl's multi-dimensional WKB quantization rule (Sec. 3.3.2)), determine the number $\bar{N}(E)$, which is the of number the eigenstates of the four-dimensional sphere with energy below or equal some energy E, reinterpret your result in terms of the Coulomb spectrum, and find the number $\bar{N}_C(E_C)$ defined as the number of the eigenstates of the Coulomb problem with energies not exceeding E_C.*

Useful information: *in four dimensions, the 3D-surface of a sphere of radius R is $S_3 = 2\pi^2 R^3$. The 4D-volume of a four-dimensional ball of radius R is $V_4 = (\pi^2/2)R^4$.*

Solution:

For a given energy E, the phase space volume occupied by the points with energy less or equal E is

$$W(\text{energy} \leq E)$$

$$= \underbrace{\text{3D-Volume(surface of 4D-sphere, cooordinates)}}_{2\pi^2 R^3}$$

$$\times \underbrace{\text{3D-Volume(3D-ball, momenta)}}_{\frac{4}{3}\pi P^3} = \frac{8\pi^3}{3} R^3 P^3 \,.$$

Thus,

$$\bar{N}(E) \overset{E \gg \hbar^2/(mR^2)}{\approx} \frac{W(\text{energy} \leq E)}{(2\pi\hbar)^3}$$

$$\overset{E \gg \hbar^2/(mR^2)}{\approx} \frac{2}{3}\left(\frac{E}{\hbar^2/(mR^2)}\right)^{3/2}$$

$$\overset{E \gg \hbar^2/(mR^2)}{\approx} \frac{1}{3}l_{4D}^3 \,.$$

Now, observe that (a) the eigenenergies of the 4D-sphere, given by Eq. 3.12, represent a monotonically increasing function of l_{4D}; (b) Coulomb eigenenergies,

$$E_{C;n,l,m} = -\frac{m\alpha^2/\hbar^2}{2n^2} \,,$$

increase monotonically with n, and (c) according to Fock's map described above, $n \overset{l_{4D} \gg 1}{\approx} l_{4D}$. Thus,

$$\bar{N}_C(E_{C;n\approx l_{4D},l,m}) \approx \bar{N}(E_{l_{4D},l_{3D},m}) \,.$$

Finally,

$$
\begin{aligned}
\bar{\mathcal{N}}_C(E_C) &\overset{E_C \gg m\alpha^2/\hbar^2}{\approx} \frac{1}{3 \times 2^{3/2}} \left(\frac{m\alpha^2/\hbar^2}{|E_C|} \right)^{3/2} \\
&\overset{E_C \gg m\alpha^2/\hbar^2}{\approx} \frac{1}{3} n^3
\end{aligned}
$$

where $E_C = -(m\alpha^2/\hbar^2)/(2n^2) = E_{C;n,l,m}$.

(b) *Using the exact quantization of the Coulomb problem, $E_{n,l,m} = -(m\alpha^2/\hbar^2)/(2n^2)$, $\{n = 1, 2, \ldots, \infty;\ l = 0, 1, \ldots, n-1;\ m = -l, -l+1, \ldots, +l\}$, find an expression for a number $\bar{\mathcal{N}}_{C;n}$ that gives the number of Coulomb eigenstates with the principal quantum number less than or equal to n; compare this number with your 4D WKB prediction from (a).*

Solution:

$$
\bar{\mathcal{N}}_{C;n} = \sum_{n'=1}^{n} \sum_{l=0}^{n'-1} \sum_{m=-l}^{+l} 1
$$

or

$$
\bar{\mathcal{N}}_{C;n} = \frac{1}{3} n(n+1/2)(n+1)
$$

The agreement with the 4D WKB prediction from (a) is indeed very good.

3.1.10 *Ground state energy of a logarithmic potential, a WKB analysis*

Consider a one-dimensional particle moving in a logarithmic potential:

$$
-\frac{\hbar^2}{2m} \frac{\partial^2}{\partial x^2} \psi(x) + 2\epsilon \ln(x/a)\psi(x) = E\psi(x) \tag{3.13}
$$

$$
x > 0. \tag{3.14}
$$

For this problem, a straightforward dimensional analysis fails to predict the ground state energy. Nevertheless, *give a dimensional estimate for the ground state energy. Use the relationship between logarithmic potentials of the same strength ϵ but different radii a to resolve the apparent ambiguity in the dimensional prediction*

Solution: Here, we have two combinations of the input parameters of a dimensionality of energy:

$$
\epsilon \text{ and } \frac{\hbar^2}{ma^2},
$$

and hence a dimensionless combination

$$P = \frac{\hbar^2}{ma^2\epsilon}.$$

A priori, the only conclusion we can make here, on dimensional grounds is that

$$E_{\text{g.s.}} \sim \epsilon f(P), \tag{3.15}$$

where we have chosen ϵ as the principal energy scale; here, $f(\xi)$ is a dimensionless function of a dimensionless variable ξ, of the order of unity.

However observe two Hamiltonians with the same ϵ but two different lengths a are different only by a constant:

$$-\frac{\hbar^2}{2m}\frac{\partial^2}{\partial x^2} + 2\epsilon \ln(x/a') = -\frac{\hbar^2}{2m}\frac{\partial^2}{\partial x^2} + 2\epsilon \ln(x/a') + 2\epsilon \ln(a/a').$$

Hence,

$$E_{\text{g.s.}}\Big|_{\epsilon, a'} = E_{\text{g.s.}}\Big|_{\epsilon, a} + 2\epsilon \ln(a/a').$$

Substitute the ansatz (3.15) to the relationship above. Assume that $a' = a + \delta a$ with $\Delta a \ll a$, Taylor-expand, and obtain $f(\xi) = \ln(\alpha \xi)$, with α being a dimensionless constant that does not depend on a. The only dimensionless function of the input parameters we have is P, but it contains a and thus can not be used to construct α. Hence α is of the order of unity. Finally,

$$\boxed{E_{\text{g.s.}} = \epsilon \ln\left(\frac{\alpha \hbar^2}{ma^2\epsilon}\right)},$$

where α is, again, an unknown constant of the order of unity. Notice the equal sign however.

3.2   Problems without provided solutions

3.2.1   *Size of a neutral meson in Schwinger's toy model of quark confinement*

Consider the one-dimensional analogue of a neutral meson in Schwinger's toy model of "color confinement":[6] one-dimensional quark and anti-quark,

---

[6] J. Schwinger, Gauge Invariance and Mass. II, *Physical Review* **128**, 2425 (1962).

of mass m each, interacting via a one-dimensional analogue of the Coulomb potential, i.e. linear potential:

$$V(x) = \alpha |x|,$$

where $x \equiv x_1 - x_2$ is the distance between the quarks. *Using the Heisenberg uncertainty relation, estimate the size of the meson in the ground state.*

3.2.2 Bohr-Sommerfeld quantization for periodic boundary conditions

Derive a Bohr-Sommerfeld quantization rule for a one-dimensional unbounded $(E > V(x))$ motion with periodic boundary conditions.

3.2.3 Ground state energy of multi-dimensional power-law potentials

Show that the dimensional estimate for the ground state energy of power-law potentials (see Problem 3.1.4) can be generalized to any number of spatial dimensions.

3.2.4 1D box as a limit of power-law potentials

Consider the one-dimensional motion of a mass m particle between two hard walls at $x = -L/2$ and $x = +L/2$.

(a) *Disregarding the fact that the exact spectrum is known, give a dimensional estimate for the spectrum of the box.*

(b) *Show that this estimate is consistent with the spectrum (3.4) of the power-law potentials, in the limit of $q \to \infty$, $K_q = A/L^{2q}$):*

$$E_n = C_q \left(\frac{\hbar^2}{m} \right)^{\frac{q}{q+1}} K_q^{\frac{1}{q+1}} (n + 1/2)^{\frac{2q}{q+1}} . \tag{3.16}$$

Do not forget that this formula is only an $n \to \infty$ limit of the exact spectrum.

(c) *Now, write down the exact solution for the spectrum of the box, and determine the $q \to \infty$ limit of the prefactor $C_\infty \equiv \lim_{q \to \infty} C_q$ above. Compare your result (which is exact) with the WKB prediction (3.5).*

(d) *Finally, using the exact solution for the harmonic oscillator, determine C_1. Compare C_1 and C_∞.*

3.2.5 *Ground state energy of a logarithmic potential, an estimate*

Consider, again, a one-dimensional particle moving in a logarithmic potential, (3.14). *Estimate the ground state energy using the Heisenberg uncertainty principle and minimization of energy.*

Answer: $E_{\text{g.s.}} \sim \epsilon \ln(\alpha \frac{\hbar^2}{ma^2\epsilon})$, where α is a dimensionless number of the order of unity.

Hint: A good idea is to replicate the scheme of the Problem 3.1.4. Unlike in the Problem 3.1.4, the kinetic and potential energies will not seem to be of the same order (and generally, the virial theorem does not directly apply to the logarithmic potential): but in reality, they are.

(b) For this problem, a straightforward dimensional analysis fails to predict the ground state energy. Nevertheless, *give a dimensional estimate for the ground state energy. Use the relationship between logarithmic potentials of the same strength ϵ but different radii a to resolve the apparent ambiguity in the dimensional prediction*

<u>Solution</u>: Here, we have two combinations of the input parameters of a dimensionality of energy:

$$\epsilon \text{ and } \frac{\hbar^2}{ma^2},$$

and hence a dimensionless combination

$$P = \frac{\hbar^2}{ma^2\epsilon}.$$

A priori, the only conclusion we can make here, on dimensional grounds is that

$$E_{\text{g.s.}} \sim \epsilon f(P), \tag{3.17}$$

where we have chosen ϵ as the principal energy scale; here, $f(\xi)$ is a dimensionless function of a dimensionless variable ξ, of the order of unity.

However observe two Hamiltonians with the same ϵ but two different lengths a are different only by a constant:

$$-\frac{\hbar^2}{2m}\frac{\partial^2}{\partial x^2} + 2\epsilon \ln(x/a') = -\frac{\hbar^2}{2m}\frac{\partial^2}{\partial x^2} + 2\epsilon \ln(x/a') + 2\epsilon \ln(a/a').$$

Hence,

$$E_{\text{g.s.}}\Big|_{\epsilon,\,a'} = E_{\text{g.s.}}\Big|_{\epsilon,\,a} + 2\epsilon \ln(a/a').$$

Substitute the ansatz (3.17) to the relationship above. Assume that $a' = a + \delta a$ with $\Delta a \ll a$, Taylor-expand, and obtain $f(\xi) = \ln(\alpha\xi)$, with α being a dimensionless constant that does not depend on a. The only dimensionless function of the input parameters we have is P, but it contains a and thus can not be used to construct α. Hence α is of the order of unity. Finally,

$$E_{\text{g.s.}} = \epsilon \ln\left(\frac{\alpha\hbar^2}{ma^2\epsilon}\right) \quad,$$

where α is, again, an unknown constant of the order of unity. Notice the equal sign however.

3.2.6 *Spectrum of a logarithmic potential*

Assuming the validity of the Bohr-Sommerfeld rule (3.1), estimate the spectrum of the logarithmic potential in (3.14). Use dimensional analysis. As in the case of the ground state energy (Problem 3.2.5), use the transformation property of the Hamiltonian under $a \to a'$.

In the Bohr-Sommerfeld quantization rule (3.1), neglect the $1/2$ correction to n.

Answer:

$$E_n = \epsilon \ln\left(\alpha\frac{\hbar^2}{ma^2\epsilon}n^2\right) \quad,$$

where α is an unknown constant of the order of unity. Notice the equal sign.

Remark: The exact WKB solution shows that $\alpha = \frac{\pi}{4}$.

3.2.7 *Closest approach to a logarithmic hill and to power-law hills*

(a) Consider motion of a one-dimensional particle in a *repulsive* logarithmic potential:

$$H(x,p) = \frac{p^2}{2m} + V(x) \quad,$$

with

$$V(x) = -\alpha\log(x/a)$$
$$\alpha > 0,\ x > 0 \quad,$$

Using dimensional analysis, show that the problem of a minimal approach to the origin for a particle that starts at infinity with an initial velocity $-v_0$ does not make sense, i.e. the minimal approach distance will always depend on the initial position, no matter how far from the origin this position is.

(b) *Using dimensional analysis, estimate the minimal approach distance for power law hills,* $V(x) = \alpha_m/x^m$, *with* $\alpha_m > 0$, $x > 0$, *and* $m = 1, 2, 3, \ldots$.

3.2.8   *Spin-1/2 in the field of a wire*

Consider the Hamiltonian for a spin-1/2 three-dimensional particle of mass M moving in the field of a straight current-carrying wire:

$$\hat{H} = -\frac{\hbar^2}{2M}\Delta - \hat{\vec{\mu}} \cdot \vec{B},$$

where the magnetic moment is given by

$$\hat{\vec{\mu}} = -g\mu_0 \hat{\vec{S}}/\hbar,$$

g is the Landé factor, μ_0 is the Bohr magneton, the wire is directed along the X-axis, and the magnetic field is given by

$$\vec{B} = \frac{2I}{cr}[\vec{e}_z \cos\Theta - \vec{e}_y \sin\Theta].$$

Here I is the current, and r and Θ are cylindrical coordinates in the $Y - Z$ plane.[7]

Using dimensional analysis, show that the ground state energy of the transverse motion can be estimated as

$$E_{g.s.} \sim \frac{Mg^2\mu_0^2 I^2}{\hbar^2 c^2}.$$

3.2.9   *Dimensional analysis of the time-dependent Schrödinger equation for a hybrid harmonic-quartic oscillator*

Consider a time-dependent version of the problem associated with the Schrödinger equation (1.1):

$$i\hbar\frac{\partial}{\partial t}\psi(x, t) = \hat{H}\psi(x, t),$$

---

[7]This problem has been solved exactly, using supersymmetric methods: see L. V. Hau, J. A. Golovchenko, and M. M. Burns, Supersymmetry and the Binding of a Magnetic Atom to a Filamentary Current, *Physical Review Letters* **74**, 3138 (1995).

with

$$\hat{H} = -\frac{\hbar^2}{2m}\frac{\partial^2}{\partial x^2} + \frac{m\omega^2}{2}x^2 + \beta x^4 \,.$$

The ground state solution reads

$$\psi_{\mathrm{g.s.}}(x,\,t) = \psi_{\mathrm{g.s.}}(x)e^{-iE_{\mathrm{g.s.}}t/\hbar}\,,$$

where $\psi_{\mathrm{g.s.}}(x)$ is a time-independent ground state of the Hamiltonian $\hat{H}$.

Show that the appearance of a new input parameter in the problem, i.e. $\hbar$, does not increase the predictive power of the dimensional analysis of the ground state energy, performed in Problem 1.1.1.

3.3 Background

3.3.1 *Bohr-Sommerfeld quantization*

All useful cases of Bohr-Sommerfeld quantization can be concisely expressed in a single formula that reads

$$\oint dx\, p(x;\, E_{\tilde{n}}) = 2\pi\hbar\left[\tilde{n} + \#\_\mathrm{soft}\_\mathrm{points} \times \Delta_{\mathrm{soft}} + \#\_\mathrm{hard}\_\mathrm{points} \times \Delta_{\mathrm{hard}}\right]$$

with $\tilde{n} = 0,\, 1,\, 2,\, 3,\, \ldots\,,$ \hfill (3.18)

where

$$\Delta_{\mathrm{soft}} = \frac{\pi}{2}$$
$$\Delta_{\mathrm{hard}} = \pi\,,$$

$\#\_\mathrm{soft}\_\mathrm{points}(\#\_\mathrm{hard}\_\mathrm{points})$ counts the number of the soft(hard) turning points encountered by the classical trajectory during the cycle of motion (see Fig. 3.3 for an example), and the phase $\Delta_{\mathrm{soft}}(\Delta_{\mathrm{hard}})$ gives the phase acquired by the wavefunction in each passage through a soft(hard) turning point. The quantum number $\tilde{n}$ counts the number of *unremovable nodes* of the wavefunction. Here, the nodes inside the classically allowed region are unremovable since they do not dissapear for small modifications of the trapping potential. On the other hand, a node at a location of a hard wall can be removed by replacing the hard wall by a very high but finite wall.

The closed-loop integral in (3.18) is understood as the classical action over a complete oscillation cycle:

$$\oint p(x|E)\,dx \equiv \int_0^{T(E)} p(t|E)v(t|E)\,dt\,,$$

where p is the canonical momentum, $v \equiv \dot{x}$ is the velocity, and T is the period of the classical motion. For the case of two turning points, the action reduces to

$$\oint p(x|E)\,dx = 2\int_{x_{min}}^{x_{max}} |p(x|E)|\,dx$$

$$p(x|E) = \pm\sqrt{2m(E - V(x))}\,.$$

There is no consistent convention on what integer the quantum number should start from. Traditionally, the following is used:

$$\text{Two soft turning points:} \quad \oint dx\,p(x;\,E) = 2\pi\hbar(n + 1/2)$$

$$\text{with} \quad n = 0,\,1,\,2,\,3,\,\ldots,$$

$$\tag{3.19}$$

$$\text{One soft and one hard turning point:} \quad \oint dx\,p(x;\,E) = 2\pi\hbar(n - 1/4)$$

$$\text{with} \quad n = 1,\,2,\,3,\,4,\,\ldots$$

$$\tag{3.20}$$

and

$$\text{Two hard turning points:} \quad \oint dx\,p(x;\,E) = 2\pi\hbar n$$

$$\text{with} \quad n = 1,\,2,\,3,\,4,\,\ldots.$$

$$\tag{3.21}$$

The first formula, (3.19), is very standard, and its derivation can be found in literally any textbook on Quantum Mechanics.[8] Derivations for the other two, (3.20) and (3.21), are harder to find. For that reason, we reproduce them in Problems 3.4.1 and 3.4.2 respectively.

3.3.2   *Multi-dimensional WKB*

Additionally, a multi-dimensional generalization of the WKB idea that each eigenstate is allocated approximately $\hbar^d$ of the phase-space volume[9] allows one to estimate the number of eigenstates, $\mathcal{N}(E)$, whose energies are less than or equal to some energy E. This number is then proportional to the phase-space volume occupied by points whose energy is less than or equal to E:

$$\bar{N}(E) \overset{E\to\infty}{\approx} \frac{1}{(2\pi\hbar)^d} \int d^d q\, d^d p\,\Theta[E - H(\vec{q},\,\vec{p})]\,,$$

---

[8]See, for example, L. D. Landau and L. M. Lifshitz, *Quantum Mechanics (Non-relativistic Theory)* (Butterworth-Heinemann, Oxford, 1981).
[9]Ibid.

where d is the number of spatial dimensions, and $\vec{q}$ and $\vec{p}$ are canonical coordinates and momenta respectively. This relationship is known as the Weyl law.[10]

3.4 Problems linked to the "Background"

3.4.1 *Bohr-Sommerfeld quantization for one soft turning point and a hard wall*

Derive the Bohr-Sommerfeld quantization rule for the case of one (left, to be specific) turning point and one hard wall (right). On one hand, the soft turning point fixes the wavefunction in the classically allowed domain to

$$\psi(x) = \frac{C_a}{\sqrt{v_c(x;\,E)}} \, \cos\left[\frac{1}{\hbar}\sigma_a(x;\,E) - \frac{\pi}{4}\right].$$

On the other hand, the right turning point demands that

$$\psi(x) = \frac{C_b}{\sqrt{v_c(x;\,E)}} \, \sin\left[\frac{1}{\hbar}\sigma_b(x;\,E)\right].$$

Reconciliation of the two gives rise to the Bohr-Sommerfeld quantization,

$$\oint dx \, p_c(x;\,E) = 2\pi\hbar(n + \delta). \qquad (3.22)$$

Here,

$$\sigma_q(x;\,E) \equiv \left| \int_q^x dx \, p_c(x;\,E) \right|$$

$$p_c(x;\,E) \equiv \sqrt{2m(E - V(x))}$$

$$v_c(x;\,E) = \frac{1}{m} \, p_c(x;\,E)$$

$$\oint dx \, p_c(x;\,E) \equiv 2 \int_a^b dx \, p_c(x;\,E).$$

Conventionally, the quantum number n in the quantization rule (3.22) starts from 0 for two soft walls, and from 1 if at least one hard wall is present. In many respects this convention is inconsistent, but it is so standard that we should not attempt to break it.

---

[10]H. Weyl, Über die asymptotische Verteilung der Eigenwerte, *Nachrichten von der Gesellschaft der Wissenschaften zu Göttingen, Mathematisch-Physikalische Klasse*, p. 110 (1911).

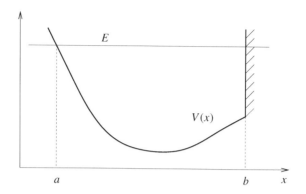

Fig. 3.3   One soft and one hard turning point.

<u>Solution:</u>

$$\psi(x) = \frac{C_a}{\sqrt{v_c(x;\,E)}} \cos\left(\frac{1}{\hbar}\sigma_a(x;\,E) - \frac{\pi}{4}\right)$$

$$= \frac{C_b}{\sqrt{v_c(x;\,E)}} \sin\left(\frac{1}{\hbar}\sigma_b(x;\,E)\right).$$

The two actions are related as

$$\sigma_b(x;\,E) = \hbar\Theta_{ab} - \sigma_a(x;\,E)\,,$$

where

$$\hbar\Theta_{ab} \equiv \int_a^b dx\, p_c(x;\,E)\,.$$

Hence

$$C_a \cos\left(\frac{1}{\hbar}\sigma_a(x;\,E) - \frac{\pi}{4}\right) = C_b \sin\left(\Theta_{ab} - \frac{1}{\hbar}\sigma_a(x;\,E)\right)$$

$$\left(= C_b \cos\left(\frac{1}{\hbar}\sigma_a(x;\,E) - \Theta_{ab} + \frac{\pi}{2}\right)\right).$$

Consider the following equation:

$$\forall x: \quad C_a \cos(f(x)) = C_b \cos(f(x) + \eta)\,,$$

where $f(x)$ is a continuous monotonic function of x. The only solutions are

$$C_a = C_b \quad \text{and} \quad \eta = 2\pi m$$

$$\text{or}$$

$$C_a = -C_b \quad \text{and} \quad \eta = 2\pi m + \pi\,,$$

where m is any integer.

The first group of solutions gives

$$\Theta_{ab} = 2\pi n' + \frac{3\pi}{4} .$$

The second gives

$$\Theta_{ab} = 2\pi n'' + \frac{7\pi}{4} .$$

Combining the two and using the fact that Θ_{ab} is a non-negative number we get

$$\Theta_{ab} = \pi \left(n - \frac{1}{4} \right)$$
$$\text{with} \quad n = 1, 2, 3, \dots ,$$

or

$$\boxed{\oint dx\, p_c(x;\, E) = 2\pi\hbar \left(n - \frac{1}{4} \right)}$$
$$\text{with} \quad n = 1, 2, 3, \dots$$

3.4.2 *Bohr-Sommerfeld quantization for two hard walls*

The same as above but for two hard walls.

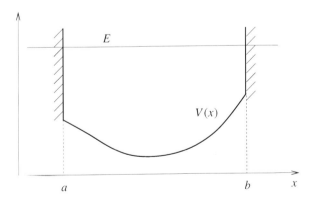

Fig. 3.4   Two hard walls.

<u>Solution:</u> Likewise:

$$C_a \sin\left(\frac{1}{\hbar}\sigma_a(x;\,E)\right) = -C_b \sin\left(\frac{1}{\hbar}\sigma_a(x;\,E) - \Theta_{ab}\right),$$

hence

$$\Theta_{ab} = \begin{cases} 2\pi n' & \text{for } C_a = C_b \\ \text{or} & \\ 2\pi n'' + \pi & \text{for } C_a = -C_b \end{cases}.$$

Combining the two and using the fact that Θ_{ab} is a non-negative number we get

$$\Theta_{ab} = \pi n$$
$$\text{with} \quad n = 1,\, 2,\, 3,\, \dots,$$

or

$$\oint dx\, p_c(x;\,E) = 2\pi\hbar n$$
$$\text{with} \quad n = 1,\, 2,\, 3,\, \dots$$

Note that we have excluded the $n = 0$ solution using yet another consideration. The classical momentum $p_c(x;\,E)$ in the definition of Θ_{ab} is a non-negative number. Therefore $\Theta_{ab} = 0$ would lead to $p_c(x;\,E) = 0$ everywhere, and thus to $\sigma_a(x;\,E) = 0$ everywhere. On the other hand, $\psi(x) \propto \sin[\frac{1}{\hbar}\sigma_a(x;\,E)]$, and thus $\psi(x) = 0$ everywhere. In this case, WKB would predict an unnormalizable solution, which must be excluded.

Chapter 4

"Halved" Harmonic Oscillator:
a Case Study

Introduction

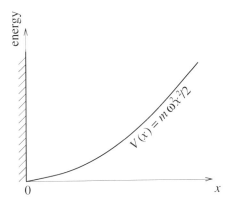

Fig. 4.1   A "halved" harmonic oscillator potential.

Consider a potential for a particle in the field of a hard wall and a harmonic force:

$$V(x) = \begin{cases} +\infty & \text{for } x < 0 \\[2mm] \dfrac{m\omega^2 x^2}{2} & \text{for } x \geq 0 \end{cases}.$$

In what follows, we will study its spectrum, using the methods described in previous Chapters.

4.1   Solved problems

4.1.1   Dimensional analysis

Apply dimensional analysis to the Bohr-Sommerfeld quantization rule and estimate the spectrum. Pay attention to the δ correction in $2\pi\hbar(n+\delta)$.

Solution: The Bohr-Sommerfeld rule reads

$$\oint dx \sqrt{2m(E_n - m\omega^2 x^2/2)} = 2\pi\hbar(n - 1/4)$$

$$n = 1, 2, 3, \ldots ,$$

where we have taken into account the fact that the left turning point is represented by a hard wall. The quantization rule above is governed by two parameters: $\eta \equiv (\hbar(n - 1/4))^2/m$ and $\Upsilon \equiv m\omega^2$. The rest of the derivation is identical to the one for the harmonic oscillator (Problem 3.1.2, sub-problem (c)). We get

$$\boxed{\begin{array}{l} E_n \sim \hbar\omega(n - 1/4) \\ n = 1, 2, 3, \ldots \end{array}}$$

4.1.2   Order-of-magnitude estimate

Make an order-of-magnitude estimate of the spectrum: use the Bohr-Sommerfeld quantization rule and approximate the phase space trajectory by a rectangle. Keep the δ correction.

Solution: The phase-space trajectory in the Bohr-Sommerfeld rule (4.1) can be approximately replaced by the "circumscribed rectangle" of the trajectory, see Fig. 4.2. The phase space integral then reads:

$$\oint dx \sqrt{2m(E_n - m\omega^2 x^2/2)} \sim \underbrace{\sqrt{2E_n/(m\omega^2)}}_{x_{max}} \underbrace{\sqrt{2mE_n}}_{p_{max}}$$

$$\sim E_n/\omega .$$

Inserting this estimate into the Bohr-Sommerfeld rule (4.1) gives

$$\boxed{\begin{array}{l} E_n \sim \hbar\omega(n - 1/4) \\ n = 1, 2, 3, \ldots \end{array}}$$

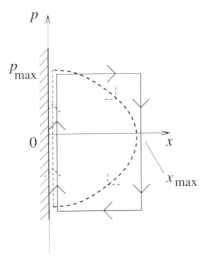

Fig. 4.2 An order-of-magnitude estimate of the WKB integral.

4.1.3 *Another order-of-magnitude estimate*

Using the quantum-classical correspondence, show that the density of energy levels of the "halved" oscillator is two times lower than the density of states of the full one.

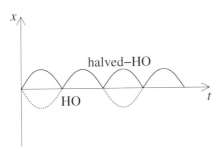

Fig. 4.3 Coordinate as a function of time, for a conventional (dashed line) and halved (solid line) harmonic oscillators.

Solution: One of the manifestations of the quantum-classical correspondence is the phenomenon that the distance between the energy levels is approximately equal to the classical frequency $\omega(E)$ multiplied by $\hbar$:

$$E_{n+1} - E_n \approx \hbar\omega((E_{n+1} - E_n)/2).$$

When we insert a wall in the middle of a harmonic oscillator, its period becomes two times shorter (see Fig. 4.3). Thus the distance between the levels increases by a factor of two as compared to the original distance $\hbar\omega$. Thus,

$$E_{n+1} - E_n \approx 2\hbar\omega$$

4.1.4 *Straightforward WKB*

Find the spectrum using the Bohr-Sommerfeld quantization rigorously.

Solution: The Bohr-Sommerfeld integral in (4.1) is the area of a half-ellipse, with radii x_{max} and p_{max} respectively (see Problem 4.1.2). We get

$$\oint dx \sqrt{2m(E_n - m\omega^2 x^2/2)} = \frac{\pi}{2} x_{max} p_{max}$$

$$= \frac{\pi}{2} \sqrt{2E_n/(m\omega^2)} \sqrt{2mE_n}$$

$$= \pi E_n/\omega.$$

Then, the Bohr-Sommerfeld rule (4.1) gives

$$E_n = 2\hbar\omega(n - 1/4)$$
$$n = 1, 2, 3, \dots$$

4.1.5 *Exact solution*

Using your knowledge about the spectrum and the properties of the eigenstates of the conventional (full) harmonic oscillator, find the spectrum of the "halved" harmonic oscillator exactly.

Solution: Two statements need to be proven.

Statement 1. An antisymmetric continuation

$$\psi(x)_{full} = \begin{cases} \psi(x)_{half} & \text{for } x \geq 0 \\ -\psi(-x)_{half} & \text{for } x < 0 \end{cases}$$

of the solution of the "halved" problem is a correct solution, including boundary conditions, of the full problem,

$$-\frac{\hbar^2}{2m}\frac{\partial^2}{\partial x^2}\psi + \frac{m\omega^2 x^2}{2}\psi = E\psi. \tag{4.1}$$

Indeed:

(i) by considering a $x \to -x$ substitution, it is easy to show that the differential equation (4.1) is obeyed not only for $x > 0$, but also for $x < 0$;
(ii) the boundary condition at $x \to -\infty$ is satisfied;
(iii) the wave function and its first derivative are continuous at $x = 0$;

Thus, each eigenstate of the "halved" problem is an odd eigenstate of the full one.

Statement 2. Every odd solution of the full problem (4.1) provides a solution of the "halved" problem in the region $x > 0$.

Indeed:

(i) it obeys the "halved" problem Schrödinger equation for $x > 0$;
(ii) it obeys both the $\psi_{\text{half}}(0) = 0$ and $\psi_{\text{half}}(+\infty) = 0$ boundary conditions;

Thus

$$E_n = E_{n'}^{\text{full}}\Big|_{n'=2n-1} = \hbar\omega(n' + 1/2)\Big|_{n'=2n-1},$$

or

$$\boxed{\begin{aligned} E_n &= 2\hbar\omega(n - 1/4) \\ n &= 1, 2, 3, \ldots \end{aligned}}$$

Chapter 5

Semi-Classical Matrix Elements of Observables and Perturbation Theory

5.1 Solved problems

5.1.1 Quantum expectation value of x^6 in a harmonic oscillator

Consider a harmonic oscillator,

$$\hat{H} = \frac{\hat{p}^2}{2m} + \frac{m\omega^2 x^2}{2}.$$

(a) *Compute the eigenstate expectation values of the sixth power of coordinate using the WKB expressions for the matrix elements.*

Useful information. For a harmonic oscillator of frequency ω:

the energy spectrum reads $E_n = \hbar\Omega(n+1/2)$;
the matrix elements of the coordinate read $x_{n,n'} = \tilde{x}(\sqrt{n}\delta_{n,n'+1} + \sqrt{n'}\delta_{n',n+1})$, where $\tilde{x} = \sqrt{\hbar/(2m\omega)}$.

Solution: Introduce a dimensionless coordinate $\xi \equiv x/\tilde{x}$, where $\tilde{x} \equiv \sqrt{\hbar/(2m\omega)}$. The matrix elements of $\hat{\xi}$ read $\xi_{n,n'} = \sqrt{n}\delta_{n,n'+1} + \sqrt{n'}\delta_{n',n+1}$. Then,

$$
\left(\hat{\xi}^6\right)_{n,n} = \sum_{n_1}\sum_{n_2}\sum_{n_3}\sum_{n_4}\sum_{n_5} \left(\sqrt{n}\delta_{n,n_1+1} + \sqrt{n_1}\delta_{n_1,n+1}\right)
$$
$$
\times \left(\sqrt{n_1}\delta_{n_1,n_2+1} + \sqrt{n_2}\delta_{n_2,n_1+1}\right)
$$
$$
\times \left(\sqrt{n_2}\delta_{n_2,n_3+1} + \sqrt{n_3}\delta_{n_3,n_2+1}\right)
$$
$$
\times \left(\sqrt{n_3}\delta_{n_3,n_4+1} + \sqrt{n_4}\delta_{n_4,n_3+1}\right)
$$
$$
\times \left(\sqrt{n_4}\delta_{n_4,n_5+1} + \sqrt{n_5}\delta_{n_5,n_4+1}\right)
$$
$$
\times \left(\sqrt{n_5}\delta_{n_5,n+1} + \sqrt{n}\delta_{n,n_5+1}\right).
$$

In a very long but straightforward calculation, the resulting twenty terms can be summed up to give

$$\left(x^6\right)_{n,\,n} = \frac{5}{2}\left(\frac{\hbar}{m\omega}\right)^3\left(n^3 + \frac{3}{2}n^2 + 2n + \frac{3}{4}\right)$$

(b) *The same as in (a) but using the WKB expressions.*
Solution: The classical trajectory reads

$$x(t) = \bar{x}\cos(\omega t)\,,$$

where the energy-dependent amplitude $\bar{x}$ is

$$\bar{x}(E) = \sqrt{\frac{2E}{m\omega^2}}\,.$$

The time average of x^6 is

$$\begin{aligned}
[x^6] &= \frac{1}{T}\int_0^T dt\,(\bar{x})^6\cos^6(2\pi t/T) \\
&= (\bar{x})^6\frac{1}{2\pi}\int_0^{2\pi} d\Theta\,\cos^6(\Theta) \\
&= \frac{5}{16}(\bar{x})^6\,.
\end{aligned}$$

Finally, substituting the classical energy E by its quantum counterpart $\hbar\omega(n+\frac{1}{2})$, we get

$$\left(x^6\right)_{n,\,n} = \frac{5}{2}\left(\frac{\hbar}{m\omega}\right)^3\left(n^3 + \frac{3}{2}n^2 + \frac{3}{4}n + \frac{1}{8}\right)$$

As expected, the WKB expression reproduces correctly the leading and subleading terms in the power-n expansion of these matrix elements.

5.1.2 *Expectation value of r^2 for a circular Coulomb orbit*

Consider the "circular orbit" eigenstates of the Coulomb problem:

$$\hat{H} = -\frac{\hbar^2}{2m}\Delta - \frac{\alpha}{r}\,.$$

where Δ is the three-dimensional Laplacian. For these states, the azimuthal quantum number l reaches its maximal possible value, $l = n - 1$, where n is the principal quantum number. Also, these states correspond to the zero

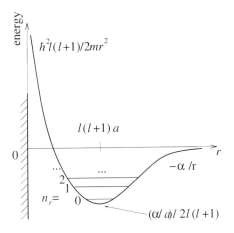

Fig. 5.1   Quantization of the radial motion in a Coulomb problem.

radial quantum number: $n_r = 0$, where $n_r \equiv n - l - 1$. The wavefunctions of the "circular orbit" eigenstates read

$$\chi_{n,\,l=n-1}(r) = a^{-1/2} \frac{2^{n+1/2}}{n^{n+1/2}\sqrt{(2n)!}} (r/a)^n \exp[-r/(na)]$$

$$\int_0^\infty dr\, |\chi_{n,\,l}(r)|^2 = 1\,,$$

where $a = \hbar^2/(m\alpha)$. Recall that the Hamiltonian is

$$\hat{H} = -\frac{\hbar^2}{2m}\Delta - \frac{\alpha}{r}\,.$$

(a) *For these states, compute the expectation value of r^2 exactly.*
   Solution: A straightforward integration gives

$$\boxed{\langle r^2 \rangle_{\text{quantim}} = n^2(n+1/2)(n+1)\,a^2}$$

(b) *For these states, compute the expectation value of r^2 using the WKB expressions. Hint: Figure 5.1 has all the information you need.*
   Solution: For the circular orbits, r does not evolve at all, being fixed at the bottom of the potential for the radial motion. Thus, $\langle r^2 \rangle_{\text{classical}} = (l(l+1)\,a)^2$. Finally

$$\langle r^2 \rangle_{\text{classical}} = (n-1)^2 n^2 \, a^2$$

(c) *Compare your result from (a) to the one from (b).*
Solution: We get

$$\langle r^2 \rangle_{\text{quantim}} = n^4 \left(1 + \frac{3}{2}\frac{1}{n} + \cdots \right) a^2$$

$$\langle r^2 \rangle_{\text{classical}} = n^4 \left(1 - \frac{2}{n} + \cdots \right) a^2 \, .$$

Thus,

$$\langle r^2 \rangle_{\text{quantim}} = \langle r^2 \rangle_{\text{classical}} \left(1 + \mathcal{O}\left(\tfrac{1}{n}\right) \right)$$

5.1.3   *WKB approximation for some integrals involving spherical harmonics*

Consider a particle sliding on the surface of a sphere. Its energy spectrum is

$$E_{l,\,m} = \frac{\hbar^2 l(l+1)}{2MR^2}$$

$$m = -l, \ -l+1, \ -l+2, \ \ldots, \ +l \, ,$$

where M is particle's mass, and R is the radius of the sphere. Recall that $\hbar m$ corresponds to the projection of the angular momentum onto the z-axis, $\hat{L}_z$; $\hbar^2 l(l+1)$ gives the square of the total momentum, $\hat{L}^2 = \hat{L}_x^2 + \hat{L}_y^2 + \hat{L}_z^2$. Here, l and m are the azimuthal and magnetic quantum numbers, respectively. The eigenstates of the system are given by the spherical harmonics:

$$|l,\,m\rangle \equiv Y_{l,\,m}(\Theta,\,\phi) \, .$$

   Using the WKB approximation for the matrix elements of operators, calculate the matrix elements of $\cos(\phi)$ between the "equatorial orbit" states, $|l,\,m = l\rangle$:

$$\langle l + \Delta l, \ m = l + \Delta l | \cos(\phi) | l, \ m = l \rangle = \ ? \, ,$$

see Figure 5.2.

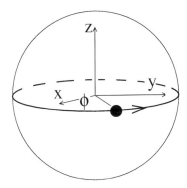

Fig. 5.2   An equatorial orbit for a particle on a sphere.

Hint: Try to match the classical time dependence of $\cos(\phi(t))$ along an equatorial orbit to its quantum time dependence:

$$\langle \psi(t) | \cos(\phi) | \psi(t) \rangle$$

$$= \sum_{l_2} \sum_{l_1} \psi^*_{l_2, m=l_2} \psi_{l_1, m=l_1} \langle l_2, m = l_2 | \cos(\phi) | l_1, m = l_1 \rangle \exp[i\omega_{l_2, l_1} t]$$

$$= \sum_{l_1} \sum_{\Delta l} \psi^*_{l_1+\Delta l, m=l_1+\Delta l} \psi_{l_1, m=l_1} \tag{5.1}$$

$$\times \langle l_1 + \Delta l, m = l_1 + \Delta l | \cos(\phi) | l_1, m = l_1 \rangle \exp[i\omega_{l_1+\Delta l, l_1} t]$$

$$\approx \sum_{\Delta l} \langle l + \Delta l, m = l + \Delta l | \cos(\phi) | l, m = l \rangle \exp[i\omega \Delta l\, t] \, ,$$

where (again, as usual) we assume that the initial state,

$$\psi(0) \rangle = \sum_l \psi_l | l, m = l \rangle \, ,$$

was a broad smooth wavepacket over the "equatorial" eigenstates, l is a typical angular momentum in the packet, $\omega_{l_2, l_1} = (E_{l_2, m=l_2} - E_{l_1, m=l_1})/\hbar$, and

$$\omega = \left. \frac{d\omega_{l+\Delta l, l}}{d\Delta l} \right|_{\Delta l=0} \approx \hbar l / (MR^2) \tag{5.2}$$

approximates a typical frequency between the neighboring states, $\omega_{l_1, l}$.

    <u>Solution:</u> For an equatorial orbit, the classical time dependence of ϕ is $\phi(t) = (L./J)t \approx \omega t$, where L is the angular momentum, $J = MR^2$ is the moment of inertia, and ω is given by (5.2). Then

$$(\cos(\phi))(t) = \cos(\omega t) = \frac{1}{2} \exp[+i\omega\, t] + \frac{1}{2} \exp[-i\omega\, t] \, .$$

Matching to (5.1) gives

$$\langle l+1, m = l+1| \cos(\phi)|l, m = l\rangle$$
$$= \langle l, m = l| \cos(\phi)|l+1, m = l+1\rangle = \frac{1}{2}$$
$$\langle l+\Delta l, m = l+\Delta l| \cos(\phi)|l, m = l\rangle \overset{\Delta l \neq \pm 1}{=} 0$$

The exact result is

$$\langle l+1, m = l+1| \cos(\phi)|l, m = l\rangle$$
$$= \langle l, m = l| \cos(\phi)|l+1, m = l+1\rangle =$$
$$= -\frac{\pi 2^{-3l-4}\sqrt{(2l+1)!(2l+3)!}(2l+1)!!}{l!((l+1)!)^2}$$
$$= -\frac{1}{2} + \frac{1}{32l^2} - \frac{5}{64l^3} + \mathcal{O}\left(\left(\frac{1}{l}\right)^4\right)$$
$$\langle l+\Delta l, m = l+\Delta l| \cos(\phi)|l, m = l\rangle \overset{\Delta l \neq \pm 1}{=} 0.$$

Remark: Recall that the sign mismatch has no significance: in the quantum case, the signs of the off-diagonal matrix elements of observables differ due to different conventions for phase factors in front of the eigenstates.

5.1.4   *Ground state wavefunction of a one-dimensional box*

Consider an infinitely deep square well,

$$V(x) = \begin{cases} 0 & \text{for } |x| < L/2 \\ +\infty & \text{for } |x| \geq L/2 \end{cases}.$$

(a) *Using dimensional analysis, estimate of the value of the ground state wavefunction at the origin, $\psi(x=0)$.*

Solution: The input parameters are $\hbar^2/m \equiv \eta$ and L. The unknown is $\psi(x=0)$. The corresponding units are $[\eta] = [\mathcal{E}][\mathcal{L}]^2$ and $[L] = [\mathcal{L}]$. Since $[\psi(x=0)] = 1/\left[\sqrt{\mathcal{L}}\right]$, we get

$$\psi(x=0) \sim 1/\sqrt{L}$$

(b) *The same as in (a) but using an order-of-magnitude estimate. For this purpose, approximate the ground state wavefunction by a piecewise linear wedge,*

$$\psi(x) \approx \psi(x = 0)(1 - 2|x|/L).$$

Solution: From the normalization we get,

$$\int_{-L/2}^{+L/2} dx \, |\psi(x = 0)|^2 (1 - 2|x|/L)^2 = 1,$$

or, up to a phase factor,

$$\psi(x = 0) \approx \sqrt{3/L}$$
$$\approx 1.73\ldots/\sqrt{L}$$

(c) *Compare your results from (a) and (b) with an exact result.*
Solution: Wave function for the ground state reads

$$\psi(x) = \sqrt{2/L} \cos(\pi x/L).$$

We get finally:

$$\psi(x = 0) \approx \sqrt{2/L}$$
$$\approx 1.41\ldots/\sqrt{L}$$

5.1.5 *Eigenstates of the harmonic oscillator at the origin: how a factor of two can restore a quantum-classical correspondence*

Consider a particle of mass m moving in a harmonic potential of frequency ω.

(a) *Find the asymptotic behavior of the probability density*

$$\rho_n(x) \equiv |\psi_n(x)|^2$$

at the origin for even eigenstates as $n \to \infty$,

$$\rho_{n=even \gg 1}(0) = ?$$

Solution: The eigenfunctions at the origin read

$$\psi_n(0) = \frac{H_n(0)}{\pi^{1/4}\sqrt{2^n n!}} \frac{1}{\sqrt{a_{\text{HO}}}}$$

$$= \begin{cases} \dfrac{(-1)^{n/2} 2^{-n/2}\sqrt{n!}}{\pi^{1/4}} \dfrac{1}{(n/2)!} \dfrac{1}{\sqrt{a_{\text{HO}}}} & \text{for } n = \text{even} \\ \\ 0 & \text{for } n = \text{odd} \end{cases},$$

where $H_n(x)$ are Hermite polynomials, and $a_{\text{HO}} \equiv \sqrt{\hbar/(m\omega)}$ is the size of the ground state. The following asymptotic can be obtained using Stirling's formula:

$$\psi_n(0) \overset{n=\text{even}\gg1}{\approx} \frac{2^{1/4}}{\sqrt{\pi}n^{1/4}} \frac{1}{\sqrt{a_{\text{HO}}}}.$$

This leads to

$$\boxed{\rho_{n=\text{even}\gg1}(0) = \frac{\sqrt{2}}{\pi\sqrt{n}} \frac{1}{a_{\text{HO}}}}$$

(b) *The same as in (a), but averaged over the quantum oscillations of density:*

$$\overline{\rho_{n\gg1}(x\approx0)}^x = ?$$

Solution: In the WKB approximation, for even eigenstates, the density in the vicinity of the origin reads

$$\rho_{n=\text{even}\gg1}(x\approx0) = \rho_{n=\text{even}\gg1}(0)\cos(\kappa(E_n)x)^2, \tag{5.3}$$

where $\kappa(E) \equiv \sqrt{2mE}/\hbar$ is the wavevector at the origin, and $E_n \approx \hbar\omega n$ is the energy as a function of the quantum number. Averaging over these quantum oscillations reduces the density by a factor of two:

$$\overline{\rho_{n=\text{even}\gg1}(x\approx0)}^x = \frac{1}{\pi\sqrt{2n}} \frac{1}{a_{\text{HO}}}.$$

For the odd states, we have

$$\rho_{n=\text{odd}\gg1}(x\approx0) = \rho_{n=\text{even}\gg1}(0)\sin(\kappa(E_n)x)^2. \tag{5.4}$$

Averaging over space leads to the same result as for the even states. Combining these two results together, we get

$$\overline{\rho_{n \gg 1}(x \approx 0)}^x = \frac{1}{\pi\sqrt{2n}}\frac{1}{a_{\text{HO}}}$$

(c) *The same as in (a), but averaged over all states, even and odd:*

$$\overline{\rho_{n \gg 1}(x \approx 0)}^n = ?$$

Solution: Since the density at the origin is zero for the odd states, averaging over quantum numbers reduces the center density by two, with respect to the value for the even states alone:

$$\overline{\rho_{n \gg 1}(x = 0)}^n = \frac{1}{\pi\sqrt{2n}}\frac{1}{a_{\text{HO}}} \; .$$

The WKB expressions (5.3,5.4) also allow one to construct the averaged over n density distribution around the origin:

$$\overline{\rho_{n \gg 1}(x \approx 0)}^n = \rho_{n=\text{even} \gg 1}(0)\frac{\cos(\kappa(E_n)x)^2 + \sin(\kappa(E_n)x)^2}{2}$$

$$= \frac{\rho_{n=\text{even} \gg 1}(0)}{2} \; .$$

$$\overline{\rho_{n \gg 1}(x \approx 0)}^n = \frac{1}{\pi\sqrt{2n}}\frac{1}{a_{\text{HO}}}$$

Remark: In particular, this result guarantees that in a thermal equilibrium, the classical (see below) and the quantum densities will be close to each other.

(d) *Now find the classical expression for the probability density at the origin as a function of energy, assuming that the particle coordinate is measured at a random time*

$$\rho_{CM, E}(0) = ?$$

Solution: Using the classical formula for probability density (5.32), we get

$$\rho_{CM, E}(0) = \frac{1}{\pi\sqrt{2E/(m\omega^2)}} \tag{5.5}$$

(e) *In your answer to sub-problem (d), replace energy by its approximate quantum value $E_n \overset{n\gg1}{\approx} \hbar\omega n$ and compare your answer to the results you obtained in (b) and (c):*

$$\rho_{\mathrm{CM},n}(0) = ?$$

Solution: Replacing E by $\hbar\omega n$ in the result of sub-problem (d) we get

$$\rho_{\mathrm{CM},n}(0) = \frac{1}{\pi\sqrt{2n}}\frac{1}{a_{\mathrm{HO}}}$$

Indeed the results of sub-problems (b), (c), and (e) coincide, producing yet another manifestation of the classical-quantum correspondence.

5.1.6 *Probability density distribution in a "straightened" harmonic oscillator*

Consider again a "straightened" harmonic oscillator, from Problem 3.1.3.
  (a) *Find the classical period of motion T.*
  Solution: Adding the propagation times for the harmonic and flat parts together, we get:

$$T(E) = T_{\mathrm{HO}} + \frac{2L}{v_0(E)},$$

where

$$T_{\mathrm{HO}} = \frac{2\pi}{\omega}$$

is the period for the harmonic oscillator, and

$$v_0(E) = \sqrt{2E/m}$$

is the velocity in the "flat" region

  (b) *For the classical motion at a given energy E find the density in the "flat" region of the potential. Be inventive: recall that the probability of finding the particle in a particular region of space is proportional to the time spent in the region.*

<u>Solution:</u> The probability of finding our particle in the region between x and $x + \Delta x$, at a random observation time is given by

$$P(x' \in [x, \, x + \Delta x]) \overset{\Delta x \to 0}{\approx} \rho_{\text{CM}}(x')\Delta x \, ,$$

where $\rho_{\text{CM}}(x)$ is the probability density, and "CM" stands for classical mechanics. On the other hand,

$$P(x' \in [x, \, x + \Delta x]) \overset{\Delta x \to 0}{\approx} \frac{\Delta t}{T} \, ,$$

where Δt is the time spent in the region $[x, \, x + \Delta x]$, and T is the period of motion. Combining all the above we get

$$\rho_{\text{CM}}(x) = \frac{2}{v(x)T} \, ,$$

where $v(x)$ is the absolute value of the classical velocity. The factor of two takes into account the fact that during a period, we pass through a given point twice.

Finally

$$\boxed{\rho_{\text{CM}}(x) = 2(v_0(E)\, T(E))^{-1}}$$

Here, $T(E)$ and $v_0(E)$ are the period and the velocity found in sub-problem (a).

(c) *In the flat region, estimate the wave function of an eigenstate of energy E.*

<u>Solution:</u> (i) While the quantum-mechanical interference patterns are inaccessible classically, the spatially averaged quantum-mechanical and classical-mechanical density distributions must agree at high energies. (ii) In the flat region the wavefunction is represented by a standing wave with a wavevector $k_0(E) \equiv \sqrt{2mE}/\hbar$. (iii) The left-right symmetry of the potential ensures that the eigenstates are either even or odd. The wavefuction that satifies all three requirements above reads

$$\boxed{\rho_{\text{QM}}(x) = 2 \, \rho_{\text{CM}}(x = 0) \times \begin{cases} \cos^2(k_0(E)x) & \text{for even states} \\ \sin^2(k_0(E)x) & \text{for odd states} \end{cases} \, ,}$$

where "QM" stands for quantum mechanics, $T(E)$ and $v_0(E)$ are the period and the velocity found in sub-problem (a), and $\rho_{CM}(x)$ is the classical density found in sub-problem (b).

5.1.7   *Eigenstates of a quartic potential at the origin*

Consider a one-dimensional particle of mass m in a

$$V(x) = \beta x^4$$

potential. Its spectrum can be estimated using the result of Problem 3.1.5; it reads

$$E_n \sim \left(\frac{\hbar^2}{m}\right)^{2/3} \beta^{1/3}\,.$$

(a) *Give an estimate for the value of the n-th eigenstate at $x = 0$:* $\psi_n(x{=}0) =?$ *Assume that n is even.*

Solution: Recall that the classical density is given by $\rho_c(x) = \frac{2}{T(E)v(x,E)}$, where $T(E)$ is the period, and $v(x, E)$ is the velocity. The value of the wave function at its crest (and for the even states we do have a crest in the middle) is of the order of the square root of the classical density:[1]

$$\psi(x) \sim \sqrt{\rho_c(x)}\,,$$

On the other hand, the density can be well estimated as

$$\rho_c \sim 1/\Delta x\,,$$

where

$$\Delta x \sim (E/\beta)^{\frac{1}{4}}\,.$$

Using the E_n dependence (5.6) we get

$$\boxed{\psi_n(x{=}0) \sim \left(\frac{m\beta}{\hbar^2}\right)^{1/12} n^{-1/6}}$$

(b) *Derive an expression (not just an order-of-magnitude estimate) that relates the de-Broglie wavelength at the origin to the energy of the state:* $\lambda_{dB}(x{=}0, E) =?$

---

[1]In fact, the square root of twice the classical density, since $|\psi(x)|^2 \approx 2\rho_c(x)\cos^2(2\pi x/\lambda_{dB})$.

Solution: By definition, the de-Broglie wavelength is

$$\lambda_{\mathrm{dB}}(x,\,E) = \frac{2\pi\hbar}{mv(x,\,E)}\,,$$

where $v(x,\,E) = \sqrt{2(E - V(x))/m}$ is the classical velocity. On the other hand

$$v(x\!=\!0,\,E) = \sqrt{2E/m}\,,$$

since $V(x\!=\!0) = 0$. Finally,

$$\lambda_{\mathrm{dB}}(x\!=\!0,\,E) = \frac{\sqrt{2}\pi\hbar}{\sqrt{mE}}$$

5.1.8 *Perturbation theory with exact and semi-classical matrix elements for a harmonic oscillator perturbed by a quartic correction or ...*

Consider a Hamiltonian for a harmonic oscillator perturbed by a quartic correction:

$$-\frac{\hbar^2}{2m}\frac{\partial^2}{\partial x^2}\psi(x) + \frac{m\omega^2}{2}x^2\psi(x) + \beta x^4\psi(x) = E\psi(x)\,.$$

Find the first order perturbation theory shift of the spectrum, $E_n^{(1)}$, using three methods:

(a) dimensional analysis applied to the Bohr-Sommefeld rule, ...
  Solution: In the first order of the perturbation theory, we get

$$E_n^{(1)} = (\hat{V})_{n,n} = \beta\left(\hat{x}^4\right)_{n,n}$$

$$= \beta\mathcal{L}^4\left(\frac{(\hbar(n+1/2))^2}{m},\, m\omega^2\right)\Phi\left(\frac{(\hbar(n+1/2))^2}{m},\, m\omega^2\right), \quad (5.6)$$

where $\mathcal{L}$ is a length scale, Φ is a dimensionless function of two dimensionful parameters, and $\frac{(\hbar(n+1/2))^2}{m}$ and $m\omega^2$ are the only two independent dimensionful parameters entering the Bohr-Sommerfeld quantization rule

$$\oint dx\,\sqrt{2m\left(E_n - \frac{m\omega^2}{2}x^2 - \beta x^4\right)} = 2\pi\hbar\left(n + \frac{1}{2}\right)$$

besides β. Observe that β itself is excluded from the list of input parameters: it has already been used to determine the order of the perturbation theory—first in this case—in Eq. (5.6). One can easily show that no dimensionless combinations can be formed out of the list, which means that $\Phi \sim 1$. Also, the only parameter with the dimension of length that can be assembled out of these two parameters is

$$\mathcal{L}(\frac{(\hbar(n+1/2))^2}{m}, m\omega^2) = \sqrt{\frac{\hbar}{m\omega}}\sqrt{n+1/2}. \qquad (5.7)$$

Thus,

$$\boxed{\begin{aligned} E_n^{(1)} &\sim \beta \left(\frac{\hbar}{m\omega}\right)^2 (n+\frac{1}{2})^2 \\ &\sim \beta \left(\frac{\hbar}{m\omega}\right)^2 (n^2 + n + \frac{1}{4}) \end{aligned}}$$

... *(b) the WKB approximation for the matrix elements of the perturbation, and ...*

   Solution: Consider now the classical trajectory

$$x(t) = \bar{x}\cos(\omega t),$$

with

$$\bar{x}(E) = \sqrt{\frac{2E}{m\omega^2}}.$$

Then, the time average of x^4 reads

$$\begin{aligned} [x^4] &= \frac{1}{T}\int_0^T dt\, (\bar{x})^4 \cos^4(2\pi t/T) \\ &= \frac{3}{8}(\bar{x})^4. \end{aligned}$$

Finally, replacing E by $\hbar\omega(n+\frac{1}{2})$, we get

$$\boxed{\begin{aligned} E_n^{(1)} &= \frac{3}{2}\beta \left(\frac{\hbar}{m\omega}\right)^2 \left(n+\frac{1}{2}\right)^2 \\ &= \frac{3}{2}\beta \left(\frac{\hbar}{m\omega}\right)^2 \left(n^2 + n + \frac{1}{4}\right) \end{aligned}}$$

... *(c) the exact solution.*

Solution: Introduce a dimensionless coordinate $\xi \equiv x/\tilde{x}$ with matrix elements $\xi_{n,n'} = \sqrt{n}\,\delta_{n,n'+1} + \sqrt{n'}\,\delta_{n',n+1}$; here, $\tilde{x} \equiv \sqrt{\hbar/(2m\omega)}$. Then,

$$
\left(\hat{\xi}^4\right)_{n,n} = \sum_{n_1}\sum_{n_2}\sum_{n_3} \left(\sqrt{n}\,\delta_{n,n_1+1} + \sqrt{n_1}\,\delta_{n_1,n+1}\right)
$$
$$
\times \left(\sqrt{n_1}\,\delta_{n_1,n_2+1} + \sqrt{n_2}\,\delta_{n_2,n_1+1}\right)
$$
$$
\times \left(\sqrt{n_2}\,\delta_{n_2,n_3+1} + \sqrt{n_3}\,\delta_{n_3,n_2+1}\right)
$$
$$
\times \left(\sqrt{n_3}\,\delta_{n_3,n+1} + \sqrt{n}\,\delta_{n,n_3+1}\right)
$$
$$
= 6n^2 + 6n + 3\,.
$$

We get finally

$$
\boxed{E_n^{(1)} = \frac{3}{2}\beta \left(\frac{\hbar}{m\omega}\right)^2 \left(n^2 + n + \frac{1}{2}\right)}
$$

5.1.9 ... or by a cubic correction

Consider a Hamiltonian for a harmonic oscillator perturbed by a cubic correction:

$$
-\frac{\hbar^2}{2m}\frac{\partial^2}{\partial x^2}\psi(x) + \frac{m\omega^2}{2}x^2\psi(x) + \alpha x^3\psi(x) = E\psi(x)\,.
$$

Similarly to Problem 5.1.8, find the second order perturbation theory shift of the spectrum, $E_n^{(2)}$, in two ways:

(a) using dimensional analysis applied to the Bohr-Sommerfeld rule and...

Solution: The second order of the perturbation theory gives

$$
E_n^{(2)} = \sum_{n'\neq n} \frac{|(\hat{V})_{n',n}|^2}{E_n^{(0)} - E_{n'}^{(0)}} = \alpha^2 \sum_{n'\neq n} \frac{|(\hat{x}^3)_{n',n}|^2}{E_n^{(0)} - E_{n'}^{(0)}}
$$
$$
= \alpha^2 \frac{\mathcal{L}^6 \left(\dfrac{(\hbar(n+1/2))^2}{m},\, m\omega^2\right)}{\mathcal{E}\left(\dfrac{(\hbar(n+1/2))^2}{m},\, m\omega^2\right)} \Phi\left(\frac{(\hbar(n+1/2))^2}{m},\, m\omega^2\right), \quad (5.8)
$$

where, again, as in Problem 5.1.8, $\mathcal{L}$ and $\mathcal{E}$ are a length scale and an energy scale respectively, Φ is a dimensionless function, and $\frac{(\hbar(n+1/2))^2}{m}$ and $m\omega^2$

are the only two independent dimensionful parameters entering the Bohr-Sommerfeld quantization rule

$$\oint dx \sqrt{2m\left(E_n - \frac{m\omega^2}{2}x^2 - \alpha x^3\right)} = 2\pi\hbar\left(n + \frac{1}{2}\right),$$

excluding the small parameter of the perturbation theory expansion, i.e. α. Since, again, no dimensionless combinations can be formed out of the former two parameters, the only available length scale is given by Eq. 5.7, and the only energy scale possible is

$$\mathcal{E}\left(\frac{(\hbar(n + 1/2))^2}{m}, m\omega^2\right) = \hbar\omega(n + 1/2),$$

the dimensional analysis unambiguously gives

$$
\begin{aligned}
E_n^{(2)} &\sim \frac{\alpha^2}{\hbar\omega}\left(\frac{\hbar}{m\omega}\right)^3\left(n + \frac{1}{2}\right)^2 \\[2mm]
&\sim \frac{\alpha^2}{\hbar\omega}\left(\frac{\hbar}{m\omega}\right)^3\left(n^2 + n + \frac{1}{4}\right)
\end{aligned}
$$

... *(b) exactly.*

Solution: The matrix elements of the third power of the dimensionless coordinate $\xi \equiv x/\tilde{x}$ (see Problems 5.1.1, 5.1.8, and alike) with $\tilde{x}$ defined as $\tilde{x} \equiv \sqrt{\hbar/(2m\omega)}$ and with matrix elements $\xi_{n,n'} = \sqrt{n}\delta_{n,n'+1} + \sqrt{n'}\delta_{n',n+1}$, read

$$\left(\hat{\xi}^3\right)_{n',n} = \begin{cases} 3\sqrt{n^3}, & \text{for } n' = n - 1 \\ 3\sqrt{(n')^3}, & \text{for } n' = n + 1 \\ \sqrt{n(n-1)(n-2)}, & \text{for } n' = n - 3 \\ \sqrt{n'(n'-1)(n'-2)}, & \text{for } n' = n + 3 \\ 0 & \text{otherwise} \end{cases} .$$

Substituting this result into the general expression (5.8), with $E_n^{(0)} = \hbar\omega(n + 1/2)$, we get finally

$$E_n^{(2)} = -\frac{15}{4}\frac{\alpha^2}{\hbar\omega}\left(\frac{\hbar}{m\omega}\right)^3\left(n^2 + n + \frac{11}{30}\right)$$

Observe that according to the known general result, the second order perturbation theory shift of the ground state energy is non-positive:

$$E_{n=0}^{(2)} = -\frac{11}{8}\frac{\alpha^2}{\hbar\omega}\left(\frac{\hbar}{m\omega}\right)^3 \le 0$$

5.1.10 *Shift of the energy of the first excited state*

Consider a harmonic oscillator weakly perturbed by a field

$$\hat{V} = \epsilon\hat{D},$$

where the operator $\hat{D}$ is given by

$$\hat{D} = \hat{p}^2/m - (m\omega x)^2,$$

$\epsilon \ll 1$ is a small parameter, ω is the frequency of the oscillator, and m is its mass.

Prove that under this perturbation, the energy shift of the first excited state is always negative.

<u>Solution:</u> Using

$$x_{n,n'} = \tilde{x}(\sqrt{n}\delta_{n,n'+1} + \sqrt{n'}\delta_{n',n+1})$$
$$p_{n,n'} = i\tilde{p}(\sqrt{n}\delta_{n,n'+1} - \sqrt{n'}\delta_{n',n+1}),$$

one obtains

$$D_{n,n'} = -\hbar\omega m\left(\sqrt{n'(n'-1)}\delta_{n,n'-2} + \sqrt{n(n-1)}\delta_{n',n-2}\right).$$

Here, $\tilde{x} = \sqrt{\hbar/(2m\omega)}$, and $\tilde{p} = m\omega\tilde{x}$. Notice that (a) $\hat{D}$ does not have diagonal matrix elements, and (b) it does not have matrix elements between the states of opposite parity. Both properties are expected semiclassically. Property (a) follows from $\hat{D}$ being proportional to the difference between kinetic and potential energies; in turn, in a harmonic oscillator, the expectation value of the kinetic energy equals the expectation value of the potential energy. (In particular, the latter follows from the virial theorem introduced in Problem 9.1.4.) Property (b) follows from the fact that $\hat{D}$ is an even function of both momentum and coordinate.

Consider the reduced Hilbert space spanned by the *odd* eigenstates, $|\text{odd}, \tilde{m}\rangle \equiv |2\tilde{m} + 1\rangle$, $\tilde{m} = 0, 1, 2, \ldots$, of the harmonic oscillator. The operator $\hat{D}$, if projected to the reduced space, manifests as a perturbation that shifts the state index by one unit, up or down. It can only couple the states characterized by the indices $\tilde{m}$ of opposite parity. Thus, odd powers of $\hat{D}$ will not have diagonal matrix elements. Therefore, the perturbation

$\hat{V}$ constitutes, within the odd-reduced Hilbert space, a pertubation whose effect is zero in the first order of perturbation theory. On the other hand, the first excited state, $|n = 1\rangle$, of the full oscillator is at the same time the ground state $|\tilde{m} = 0\rangle$ of the "odd" oscillator. Then, the non-positivity of the second order perturbation theory shift of the ground state applies to the $|n = 1\rangle$ state as well.

$$\boxed{Q.E.D.}$$

Remark: A generalization of the result above is as follows. Consider a one-dimensional Hamiltonian

$$\hat{H}_0 = \frac{\hat{p}^2}{2m} + V(x)$$

and a perturbation over it,

$$\delta V(x).$$

Assume that both $V(x)$ and $\delta V(x)$ are even:

$$V(-x) = V(x)$$
$$\delta V(-x) = \delta V(x).$$

Assume also that (as usual) the unperturbed ground state is even and the unperturbed first excited state is odd:

$$\psi_{n=0}^{(0)}(-x) = \psi_{n=0}^{(0)}(x)$$
$$\psi_{n=1}^{(0)}(-x) = -\psi_{n=1}^{(0)}(x).$$

Under these conditions, the second order perturbation theory correction to the energy of the *first excited state* is always non-positive:

$$E_{n=1}^{(2)} \leq 0.$$

5.1.11 *Impossible potentials*

Consider a single one-dimensional particle moving between two hard walls, one of which is at the origin, separated by a distance L. Someone adds, next to the origin, a localized perturbation that is unknown to you. The only information you have is that the effect of the perturbation on the energy levels can be well modeled by an unusual boundary condition at the origin. The boundary condition reads:

$$\psi'(x=0) = -\psi(x=0)/a, \tag{5.9}$$

where a is a known constant length. Note that if a is moved to zero, the perturbation has no effect at all. Indeed, in this case, the boundary condition above will be simply reduced to a condition $\psi(x=0) = 0$ that is the same as the one imposed by the hard wall, already present. It is now tempting to conjecture that the length a is a prefactor in front of a localized perturbation.

Nevertheless, prove that the perturbation can not be represented by a potential independent of a with a as a prefactor in front of it:

$$\hat{H} \neq -\frac{\hbar^2}{2m} \frac{\partial^2}{\partial x^2} + \frac{a}{L} V(x), \tag{5.10}$$

where $1/L$ is there for dimensional reasons.

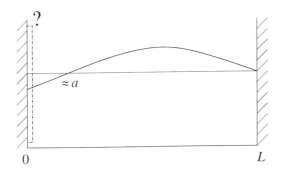

Fig. 5.3   Unknown perturbation.

<u>Useful information.</u> Note that the length a corresponds to the first node of the wavefunction only approximately, in the small a limit. Figure 5.3 can not distinguish between the node and the point a, due to the finite line thickness.

<u>Useful information.</u> If you are on the right track, you will arrive at the following transcendental equation:

$$\tan(f(\epsilon)) = A\,f(\epsilon)\,\epsilon + B\,\epsilon$$
$$f(\epsilon) \overset{\epsilon \to 0}{\to} 0.$$

The Taylor expansion of its solution reads

$$f(\epsilon) = B\epsilon + AB\epsilon^2 + \mathcal{O}(\epsilon^3).$$

<u>Hint.</u> One of the ways to solve the problem is to use the non-positivity of the second order perturbation theory shift of the ground state energy.

Useful information. This problem comes from real life. The condition (5.9) is a very typical example of scattering data, from an unknown scatterer.

Solution: Let k be the wavevector corresponding to the ground state energy:

$$E_{g.s.}(a) = \frac{\hbar^2 [k_{g.s.}(a)]^2}{2m} .$$

To satisfy the boundary condition at $x = L$ we demand

$$\psi_{g.s.}(x) = \text{const} \times \sin[k_{g.s.}(a)(x - L)]$$

for the ground state wavefunction $\psi_{g.s.}(x)$. The condition (5.9) at $x = 0$ gives

$$\tan[k_{g.s.}(a) L] = k_{g.s.}(a) a . \tag{5.11}$$

The intemediate goal is to solve approximately the transcedental equation above, in the vicinity of the unperturbed ground state wavevector

$$k_{g.s.}(a{=}0) = \frac{\pi}{L} .$$

Introduce a set of useful notations:

$$\delta \equiv \frac{a}{L}$$

$$\xi(\delta) \equiv (k_{g.s.}(a) - k_{g.s.}(0))L = k_{g.s.}(a)L - \pi .$$

The Eq. (5.11) becomes

$$\tan[\xi(\delta)] = \delta(\xi(\delta) + \pi) \tag{5.12}$$

$$\xi(\delta) \overset{\delta \to 0}{\to} 0 .$$

Expand $\xi(\delta)$ into a Taylor series:

$$\xi(\delta) = a\delta + b\delta^2 + \mathcal{O}(\delta^3) ,$$

where a and b are the unknown coefficients to be determined. The left hand side of the Eq. (5.12) becomes

$$\tan[\xi(\delta)] = \xi(\delta) + \frac{1}{3}\xi(\delta)^3 + \mathcal{O}(\xi(\delta)^5)$$

$$= (a\delta + b\delta^2 + \mathcal{O}(\delta^3)) + \frac{1}{3}(a\delta + b\delta^2 + \mathcal{O}(\delta^3))^3 + \mathcal{O}(\delta^5)$$

$$= a\delta + b\delta^2 + \mathcal{O}(\delta^3) .$$

The right hand side gives

$$\delta(\xi(\delta) + \pi) = \delta((a\delta + b\delta^2 + \mathcal{O}(\delta^3)) + \pi)$$

$$= \pi\delta + a\delta^2 + \mathcal{O}(\delta^3) .$$

We get

$$a\delta + b\delta^2 + \mathcal{O}(\delta^3) = \pi\delta + a\delta^2 + \mathcal{O}(\delta^3) \,.$$

Solving the above we obtain

$$\xi(\delta) = \pi\delta + \pi\delta^2 + \mathcal{O}(\delta^3) \,,$$

or

$$k_{g.s.}(a) = \frac{\pi}{L}\left(1 + \frac{a}{L} + \left(\frac{a}{L}\right)^2 + \mathcal{O}\left(\left(\frac{a}{L}\right)^3\right)\right).$$

For the ground state energy we get

$$E_{g.s.}(a) = \frac{\hbar^2}{2m}\left(\frac{\pi}{L}\right)^2\left(1 + 2\left(\frac{a}{L}\right) + 3\left(\frac{a}{L}\right)^2 + \mathcal{O}\left(\left(\frac{a}{L}\right)^3\right)\right)$$

Notice that the alleged second order perturbation theory correction to the energy is positive. This proves that the unknown perturbation does not have the form given in the right hand side of the inequality (5.10).

$$\boxed{Q.E.D.}$$

An example of a perturbation that generates the boundary condition (5.9) is a very narrow, very deep potential well next to the left wall whose ground state energy is very close to zero. In this case the length a will be proportional to this ground state energy. Another example of an object that produces the condition (5.9) is a short-range interaction between two fermions in a transversally cold waveguide.[2]

5.1.12 *Correction to the frequency of a harmonic oscillator as a perturbation*

Consider two harmonic oscillators

$$\hat{H}_0 = \frac{\hat{p}^2}{2m} + \frac{m\omega_0^2 x^2}{2}$$

$$\hat{H} = \frac{\hat{p}^2}{2m} + \frac{m(1+\delta)\omega_0^2 x^2}{2}$$

with frequencies ω_0 and $\sqrt{1+\delta}\,\omega_0$ respectively.

---

[2]B. E. Granger and D. Blume, Tuning the interactions of spin-polarized fermions using quasi-one-dimensional confinement, *Physical Review Letters* **92**, 133202 (2004).

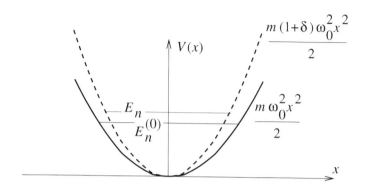

Fig. 5.4   Correction to frequency as a preturbation.

(a) *Considering the difference between the Hamiltonians,*

$$\hat{V} \equiv \hat{H} - \hat{H}_0 = \delta \frac{m\omega_0^2 x^2}{2},$$

as a perturbation of $\hat{H}_0$, find the first and the second perturbation theory corrections to the energy $E_n^{(0)} = \hbar\omega_0(n+1/2)$ of the n-th eigenstate.

<u>Useful information.</u> For a harmonic oscillator of frequency Ω:

the energy spectrum reads $E_n = \hbar\omega(n+1/2)$;
matrix elements of the coordinate read $x_{n,n'} = \tilde{x}(\sqrt{n}\delta_{n,n'+1} + \sqrt{n'}\delta_{n',n+1})$, where $\tilde{x} = \sqrt{\hbar/(2m\Omega)}$.

<u>Solution:</u> The matrix elements of $\hat{x}^2$:

$$(\hat{x}^2)_{n,n''}/\tilde{x}^2 = \sum_{n'}(\sqrt{n}\delta_{n,n'+1} + \sqrt{n'}\delta_{n',n+1})$$

$$\times(\sqrt{n'}\delta_{n',n''+1} + \sqrt{n''}\delta_{n'',n'+1})$$

$$= \sum_{n'}(\sqrt{n}\delta_{n,n'+1} + \sqrt{n+1}\delta_{n',n+1})$$

$$\times(\sqrt{n''+1}\delta_{n',n''+1} + \sqrt{n''}\delta_{n'',n'+1})$$

$$= \sqrt{n}\sqrt{n''+1}\delta_{n,n''+2} + \sqrt{n+1}\sqrt{n''+1}\delta_{n''+1,n+1}$$

$$+ \sqrt{n}\sqrt{n''}\delta_{n,n''} + \sqrt{n+1}\sqrt{n''}\delta_{n''-1,n+1}$$

$$= \begin{cases} 2n+1 & \text{for } n'' = n \\ \sqrt{n+1}\sqrt{n+2} & \text{for } n'' = n+2 \\ \sqrt{n-1}\sqrt{n} & \text{for } n'' = n-2 \\ 0 & \text{otherwise} \end{cases},$$

where $\tilde{x} = \sqrt{\hbar/(2m\omega_0)}$.

Then, for the first order correction we get:

$$E_n^{(1)} = \langle n|\hat{V}|n \rangle$$
$$= \frac{\delta m \omega_0^2}{2} (\hat{x}^2)_{n,n} .$$

Finally:

$$\boxed{E_n^{(1)} = \frac{n + \frac{1}{2}}{2} \delta \hbar \omega_0}$$

For the second order:

$$E_n^{(2)} = \frac{|\langle n|\hat{V}|n-2 \rangle|^2}{E_n^{(0)} - E_{n-2}^{(0)}} + \frac{|\langle n|\hat{V}|n+2 \rangle|^2}{E_n^{(0)} - E_{n+2}^{(0)}}$$
$$= \left(\frac{\delta m \omega_0^2}{2} \right)^2 \left\{ \frac{\tilde{x}^2(n-1)n}{2\hbar\omega_0} - \frac{\tilde{x}^2(n+1)(n+2)}{2\hbar\omega_0} \right\} .$$

Finally:

$$\boxed{E_n^{(2)} = -\frac{n + \frac{1}{2}}{8} \delta^2 \hbar \omega_0}$$

(b) *Find the exact perturbed energy E_n. Compare your findings in sub-problem (a) with the exact result. In other words, expand the exact energy in powers of δ and compare it with the results of the perturbation theory.*

Solution: The spectrum of $\hat{H}$ is

$$E_n = \hbar\omega \left(n + \frac{1}{2} \right),$$

where

$$\omega = \sqrt{1+\delta}\,\omega_0 = \left(1 + \frac{\delta}{2} - \frac{\delta^2}{8} + \cdots \right) \omega_0 .$$

The Taylor expansion (in powers of δ) of the exact eigenenergies E_n gives the same result as the perturbation theory:

$$\boxed{\begin{aligned} E_n &= E_n^{(0)} + E_n^{(1)} + E_n^{(2)} + \cdots \\ &= \hbar\omega_0 \left(n + \frac{1}{2} \right) + \frac{n + \frac{1}{2}}{2} \delta\hbar\omega_0 - \frac{n + \frac{1}{2}}{8} \delta^2 \hbar\omega_0 + \cdots \end{aligned}}$$

5.1.13   *Outer orbital of sodium atom*

The sodium atom $_{11}$Na has eleven electrons (charge $-|e|$) interacting with a nucleus (charge $Z|e|$) and between themselves. Ten inner electrons (two on the $1s$ orbital, two on the $2s$, six on the $2p$ form a nearly spherically symmetric bubble that screens the potential of the nucleus. The eleventh, outer, electron is going to reside on one of the three orbitals: $3s$, $3p$, or $3d$. The question is: which of the three will it prefer?

To answer this question, the following crude model applies:

(a) close to the nucleus, we will neglect the screening completely;
(b) far away from the nucleus, we will assume full screening;
(c) we will make the transition between these two regimes sharp. It will happen abruptly (Fig. 5.5), at a Thomas-Fermi "half-cloud" radius from the nucleus:

$$V(r) = \begin{cases} -\frac{Ze^2}{r}, & 0 < r < R_{TF} \\ -\frac{e^2}{r}, & R_{TF} < r < \infty \end{cases} . \qquad (5.13)$$

Here $|e|$ is the absolute value of the electron charge, Z is the number of protons in the nucleus. The Thomas-Fermi "half-cloud" radius is the Thomas-Fermi theory prediction (see Chapter 10) for a radius within which exactly one half of the electrons are situated:

$$R_{TF} = \eta a_B / Z^{1/3} .$$

Here, $a_B = \hbar^2/(me^2)$ is the Bohr radius, m is the electron mass, and $\eta = 1.33\ldots$.

The assignment is: using the difference between the potential (5.13) and the outer range potential $-e^2/r$,

$$\delta V(r) = \begin{cases} -\frac{(Z-1)e^2}{r}, & 0 < r < R_{TF} \\ 0, & R_{TF} < r < \infty \end{cases} ,$$

as a perturbation, estimate the correction to the energy caused by δV. Compare three situations corresponding to the unperturbed state belonging to s ($n = 3$, $l = 0$), p ($n = 3$, $l = 1$), or d ($n = 3$, $l = 2$) orbital. Note that all three have the same unperturbed energy $E_{H;\,n=3} = (me^2/\hbar^2)/18$ (generally, $E_{H;\,n} = (me^2/\hbar^2)/(2n^2)$). The unperturbed eigenstates of $-e^2/r$ (plus the corresponding centifugal potential $\hbar^2 l(l+1)/2mr^2$) for the s, p, and d orbitals read:

$$\chi_{n=3,\,l}(r) = \frac{1}{\sqrt{a_B}} f_{n=3,\,l}(r/a_B),$$

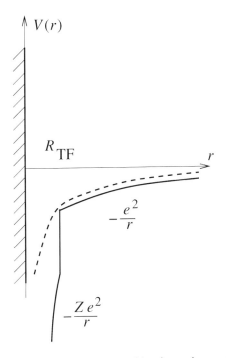

Fig. 5.5   A crude model for the potential created by the nucleus and the inner electrons.

where

$$f_{n=3,\,l=0}(x) = \frac{2}{3\sqrt{3}}x \left(1 - \frac{2}{3}x + \frac{2}{27}x^2\right) e^{-x/3}$$

$$f_{n=3,\,l=1}(x) = \frac{8}{27\sqrt{6}}x^2 \left(1 - \frac{1}{6}x\right) e^{-x/3} \qquad (5.14)$$

$$f_{n=3,\,l=2}(x) = \frac{4}{81\sqrt{30}}x^3 e^{-x/3}$$

Compare the results and choose the best candidate for the outer orbital of the ground state of the sodium atom. The states (5.14) are normalized in such a way that the matrix elements of a radially symmetric function $A(r)$ will read

$$\langle n,\, l|A(r)|n',\, l'\rangle = \delta_{l,l'} \int_0^\infty \chi^\star_{n,l} A(r)\chi_{n',l}\,.$$

Note that the $\delta_{l,l'}$ comes from the orthogonality of the angular parts of the eigenstates.

In your calculations, assume that the nuclear charge is very large, $Z \to \infty$. Retain only the dominant (highest power of Z) term. Hint: in this limit, the Thomas-Fermi radius R_{TF} becomes much smaller than the spatial extent a_B of the wavefunctions, so that only the limiting behavior of the wavefunction at zero is essential.

<u>Solution:</u> The dominant behavior of $\chi_{n,l}(r)$ at small distances is given by

$$\chi_{n,l}(r) \overset{r \to 0}{\sim} (a_B)^{-1/2}(r/a_B)^{\alpha(l)} ,$$

where

$$\alpha(l) = l + 1 .$$

The first order perturbation theory correction to the hydrogen energies $E_n^H = (me^2/\hbar^2)/(2n^2)$ thus becomes

$$
\begin{aligned}
\Delta E_{n,l} &= \langle n, l | \delta V | n, l \rangle \\
&= \int_0^\infty \delta V(r) |\chi_{n,l}(r)|^2 \\
&\sim -Z(e^2/a_B)(a_B)^{-2\alpha(l)} \int_0^{R_{TF}} r^{2l+1}\, dr \\
&\sim -\frac{E_1^H}{Z^{\frac{2l-1}{3}}} ,
\end{aligned}
$$

where E_1^H is the absolute value of the ground state energy of the hydrogen atom, also known as Rydberg. Also, $1\,\mathrm{Ry} = 13.60569\,\mathrm{eV}$. In the large atom limit, $Z \to \infty$ the s-orbital ($l = 0$) has the lowest energy.

Finally,

$$\boxed{\Delta E_{n,l} \sim -\frac{E_1^H}{Z^{\frac{2l-1}{3}}}}$$

$$\boxed{\text{the ground state orbital for } {}_{11}\mathrm{Na} \text{ is } 3s}$$

Fig. 5.6 shows the empirical sodium spectrum, in comparison with the the hydrogen spectrum. We can see that indeed, the effect of "imperfect

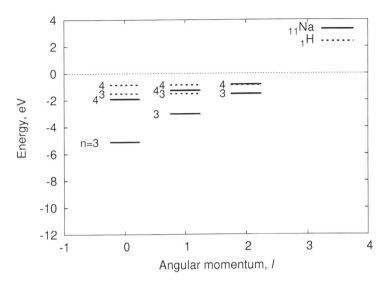

Fig. 5.6 This diagram compares the empirical energy levels of sodium with the ones of hydrogen. For high values of the angular momentum l, the sodium levels are close to the hydrogen ones. For low l, however, the effect of the unscreened nucleus becomes prominent, and the hydrogen spectrum can no longer be used as a model for sodium energy levels.

screening" increases for low angular momenta l. On the other hand, starting from $l = 3$, the sodium spectum is indistinct from the hydrogen one.

One can go further and extract the actual numerical values of the corrections. Instead of referring to the particular formulas for the $n = 3$ case, we will use a more general expression for the short-range behavior of $\chi_{n,l}$:

$$\chi_{n,l}(r) \overset{r \to 0}{\sim} (a_{\mathrm{B}})^{-1/2} \frac{2^{l+1}}{n^{l+2}(2l+1)!} \sqrt{\frac{(n+l)!}{(n-l-1)!}} (r/a_{\mathrm{B}})^{l+1} \,,$$

We get

$$\Delta E_{n,l} = -C_{n,l} \frac{E_1^H}{Z^{\frac{2l-1}{3}}}$$

$$C_{n,l} = \frac{4^{l+1} n^{-2(l+2)} (n+l)! \eta^{2(l+1)}}{(l+1)(2l+1)!^2 (n-l-1)!}$$

In particular, these formulas predict the sodium ground state energy to be lower than that of hydrogen. The value of this shift is estimated to be equal to

$$\Delta E_{n=3,\,l=3} = -7.90\,\text{eV}\,.$$

At the same time, the *empirical* value of this shift is

$$\Delta E_{n=3,\,l=3} \equiv E_{\text{Na};\,n=3,\,l=3} - E_{\text{H};\,n=3,\,l=3} = -3.61\,\text{eV}\,.$$

Even though the "imperfect screening" energy shift is by no means smaller than the hydrogen energy $E_{\text{H};\,n=3,\,l=3} = -1.51\,\text{eV}$, our perturbative estimate is only a factor of two off the exact value.

5.1.14   Relative contributions of the expectation values of the unperturbed Hamiltonian and the perturbation to the first and the second order perturbation theory correction to energy

Consider a Hamiltonian

$$\hat{H} = \hat{H}_0 + \epsilon \hat{V}$$

that consists of an unperturbed part $\hat{H}_0$ and a perturbation $\hat{V}$. Consider its exact eigenenergy E_n. The coefficients $E_n^{(m)}$ in the perturbative expansion

$$E_n = E_n^{(0)} + \epsilon E_n^{(1)} + \epsilon^2 E_n^{(2)} + \cdots$$

of E_n are the "holy grail" of perturbation theory.

On the other hand, the eigenenergy E_n can be formally written as a sum of two contributions: one coming from the unperturbed part and another from the perturbation,

$$E_n = \langle \psi_n | \hat{H} | \psi_n \rangle = \langle \psi_n | \hat{H}_0 | \psi_n \rangle + \langle \psi_n | \hat{V} | \psi_n \rangle\,.$$

Each of the two contributions can be expanded onto a power series with respect to ϵ:

$$\langle \psi_n | \hat{H}_0 | \psi_n \rangle = \langle \psi_n | \hat{H}_0 | \psi_n \rangle^{(0)} + \epsilon \langle \psi_n | \hat{H}_0 | \psi_n \rangle^{(1)} + \epsilon^2 \langle \psi_n | \hat{H}_0 | \psi_n \rangle^{(2)} + \cdots$$
$$\langle \psi_n | \hat{V} | \psi_n \rangle = \epsilon \langle \psi_n | \hat{V} | \psi_n \rangle^{(1)} + \epsilon^2 \langle \psi_n | \hat{V} | \psi_n \rangle^{(2)} + \cdots\,.$$

Finally, the corresponding orders in the expansion of E_n can be represented as a sum of two terms, one originating from $\hat{H}_0$ and another from $\hat{V}$:

$$E_n^{(0)} = \langle \psi_n | \hat{H}_0 | \psi_n \rangle^{(0)}$$
$$E_n^{(1)} = \langle \psi_n | \hat{H}_0 | \psi_n \rangle^{(1)} + \epsilon \langle \psi_n | \hat{V} | \psi_n \rangle^{(1)}$$
$$E_n^{(2)} = \langle \psi_n | \hat{H}_0 | \psi_n \rangle^{(2)} + \epsilon \langle \psi_n | \hat{V} | \psi_n \rangle^{(2)}$$
$$\cdots\,.$$

Find the relative contributions of the unperturbed part $\hat{H}_0$ and perturbation $\hat{V}$ in the first and second perturbative orders of E_n.
<u>Solution:</u> To the first order, E_n reads

$$E_n^{(1)} = \langle \psi_n^{(0)} | \hat{V} | \psi_n^{(0)} \rangle \tag{5.15}$$

To the same order, the expectation value of the unperturbed part vanishes:

$$\begin{aligned}
\langle \psi_n | \hat{H}_0 | \psi_n \rangle^{(1)} &= \langle \psi_n^{(1)} | \hat{H}_0 | \psi_n^{(0)} \rangle + \langle \psi_n^{(0)} | \hat{H}_0 | \psi_n^{(1)} \rangle \tag{5.16} \\
&= \langle \phi_n^{(1)} | \hat{H}_0 | \phi_n^{(0)} \rangle + \langle \phi_n^{(0)} | \hat{H}_0 | \phi_n^{(1)} \rangle \\
&= E_n^{(0)} \left(\langle \phi_n^{(1)} | \phi_n^{(0)} \rangle + \langle \phi_n^{(0)} | \phi_n^{(1)} \rangle \right) \\
&= 0
\end{aligned}$$

Here and below we use the known perturbation theory results and standard definitions of the intermediate objects thereof (see Eqs. 5.35, 5.34, and 5.36 below).

The first order of the perturbation gives the following intuitively expected result:

$$\begin{aligned}
\langle \psi_n | \hat{V} | \psi_n \rangle^{(1)} &= \langle \psi_n^{(0)} | \hat{V} | \psi_n^{(0)} \rangle \tag{5.17} \\
&= E_n^{(1)}
\end{aligned}$$

In summary we get

$$E_n^{(1)} = \langle \psi_n | \hat{H}_0 | \psi_n \rangle^{(1)} + \langle \psi_n | \hat{V} | \psi_n \rangle^{(1)} \,,$$

where

$$\langle \psi_n | \hat{H}_0 | \psi_n \rangle^{(1)} = 0$$
$$\langle \psi_n | \hat{V} | \psi_n \rangle^{(1)} = E_n^{(1)}$$

In the second order, we have

$$E_n^{(2)} = \sum_{n' \neq n} \frac{|\langle \psi_{n'}^{(0)} | \hat{V} | \psi_n^{(0)} \rangle|^2}{E_n^{(0)} - E_{n'}^{(0)}} \,.$$

The contribution from $\hat{H}_0$ is

$$
\begin{aligned}
\langle\psi_n|\hat{H}_0|\psi_n\rangle^{(2)} &= \langle\psi_n^{(2)}|\hat{H}_0|\psi_n^{(0)}\rangle + \langle\psi_n^{(1)}|\hat{H}_0|\psi_n^{(1)}\rangle + \langle\psi_n^{(0)}|\hat{H}_0|\psi_n^{(2)}\rangle \\
&= 2A_n^{(2)}\langle\phi_n^{(0)}|\hat{H}_0|\phi_n^{(0)}\rangle + \langle\psi_n^{(1)}|\hat{H}_0|\psi_n^{(1)}\rangle \\
&\qquad + \langle\phi_n^{(2)}|\hat{H}_0|\phi_n^{(0)}\rangle + \langle\phi_n^{(2)}|\hat{H}_0|\phi_n^{(0)}\rangle \\
&= 2A_n^{(2)}E_n^{(0)} + \langle\psi_n^{(1)}|\hat{H}_0|\psi_n^{(1)}\rangle \\
&\qquad + E_n^{(0)}\left(\langle\phi_n^{(2)}|\phi_n^{(0)}\rangle + \langle\phi_n^{(0)}|\phi_n^{(2)}\rangle\right) \\
&= 2A_n^{(2)}E_n^{(0)} + \langle\psi_n^{(1)}|\hat{H}_0|\psi_n^{(1)}\rangle \\
&\;\;\cdots \\
&= -E_n^{(2)}\,, \tag{5.18}
\end{aligned}
$$

where again we used the known results (Eqs. 5.35, 5.34, and 5.36 below). Interestingly, the second order perturbation theory correction to the normalization factor A_n contributes to this relationship.

Likewise,

$$
\begin{aligned}
\langle\psi_n|\hat{V}|\psi_n\rangle^{(2)} &= \langle\psi_n^{(1)}|\hat{V}|\psi_n^{(0)}\rangle + \langle\psi_n^{(0)}|\hat{V}|\psi_n^{(1)}\rangle \tag{5.19} \\
&= 2E_n^{(2)}
\end{aligned}
$$

In the second order, we get

$$
\boxed{
\begin{aligned}
&E_n^{(2)} = \langle\psi_n|\hat{H}_0|\psi_n\rangle^{(2)} + \langle\psi_n|\hat{V}|\psi_n\rangle^{(2)}\,, \\
&\text{where} \\
&\langle\psi_n|\hat{H}_0|\psi_n\rangle^{(2)} = -E_n^{(2)} \\
&\langle\psi_n|\hat{V}|\psi_n\rangle^{(2)} = 2E_n^{(2)}
\end{aligned}
}
$$

5.2 Problems without provided solutions

5.2.1 *A perturbation theory estimate*

Consider a one-dimensional quantum particle moving in an infinitely deep square well of width L:

$$
V(x) = \begin{cases} 0 & \text{for } |x| \le L/2 \\ +\infty & \text{for } |x| > L/2 \end{cases}. \tag{5.20}
$$

Add a small potential bump of width a in the middle:

$$\delta V(x) = \begin{cases} V_0 & \text{for } |x| \leq a/2 \\ 0 & \text{for } |x| > a/2 \end{cases} .$$

Estimate the correction to the eigenenergies caused by the bump. Assume that for the eigenstates considered, the particle momentum is high enough to treat the bump using WKB:

$$|p| \gg \hbar/a .$$

5.2.2 *Eigenstates of a two-dimensional harmonic oscillator at the origin*

Consider a Schrödinger equation for a particle of mass m moving in the field of a two-dimensional harmonic potential:

$$-\frac{\hbar^2 \Delta_{\vec{r}}}{2m}\psi(\vec{r}) + \frac{m\omega^2 r^2}{2}\psi(\vec{r}) = E\psi(\vec{r})$$

with a frequency ω, where $\vec{r} = (x, y)$ is a two-dimensional coordinate, and $\Delta_{\vec{r}}$ is the two-dimensional Laplacian. In what follows, we will be studying the probability density in the center of the trap, for the zero angular momentum eigenstates $\psi_{n_r, m=0}(r, \phi) \equiv \phi_{n_r}(r)$. The exact quantum result reads

$$|\phi_{n_r}(r = 0)|^2 = \frac{1}{\pi a_{\text{HO}}^2} : \qquad (5.21)$$

the density is *the same* for all cylindrically symmetric eigenstates. Here,

$$r = \sqrt{x^2 + y^2}$$
$$\phi = \arg(x, y)$$

are the cylindrical coordinates, and

$$a_{\text{HO}} \equiv \sqrt{\frac{\hbar}{m\omega}}$$

is the size of the ground state, $n_r = 0$, $m = 0$. The full spectrum of a 2D oscillator reads

$$E_{n_r, m} = \hbar\omega(2n_r + |m| + 1)$$
$$n_r = 0, 1, 2, \ldots$$
$$m = 0, \pm 1, \pm 2, \ldots ,$$

where n_r is the radial quantum number, and m is the projection of the angular momentum onto the Z-axis.

Observe that, superficially, the result (5.21) seems purely quantum: there is a Plank's constant in the denominator and no quantum numbers to complete it to a classical canonical action. However ...

... (a) Consider a collection of classical zero-angular-momentum trajectories—each of which is simply a straight line segment drawn through the origin—each at energy E (see Fig. 5.7). Assume that the angle between the trajectory and the X-axis is chosen at random, according to a uniform distribution. Assume further that the particle positions are later detected, at a random time. *Compute the classical probability density distribution. Show that this distribution diverges at the origin;*

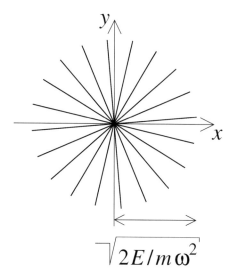

Fig. 5.7   A bundle of classical trajectories associated with an energy E zero-angular-momentum eigenstate of a two-dimensional harmonic oscillator.

(b) *For a given energy E, estimate the de-Broglie wavelength, $\lambda_{dB}(E)$;*

(c) *Assume that the quantum-mechanical density (5.21) remains approximately equal to its peak value (5.21) within a circle of radius $\lambda_{dB}(E)$ with a center at the origin. Observe that in this case, the average of the density over this disk is also (trivially) comparable to the value (5.21);*

(d) *Average the classical density over the same disc (assume its radius is much smaller than the classical amplitude of motion) and compare your result with that from (c).*

5.2.3 *Approximate WKB expressions for matrix elements of observables in a harmonic oscillator*

For a harmonic oscillator of frequency ω and mass M, find the magnitudes of the matrix elements of

- *(a)* p^3 ;
- *(b)* xp ;
- *(c)* x^4 ,

in the WKB approximation. Use solution (5.31) as an example.

5.2.4 *Off-diagonal matrix elements of the spatial coordinate for a particle in a box*

Consider again a one-dimensional quantum particle moving in an infinitely deep square well of length L (see (5.20)). *Find the semi-classical expressions for the absolute values of the matrix elements of the spatial coordinate between neighboring eigenstates:*

$$|x_{n+1,n}| = ?.$$

This is a rare example where the semi-classical approximation is more complicated than the exact solution.

5.2.5 *Harmonic oscillator perturbed by a δ-potential, ...*

Using the semiclassical expressions for density (see subsection (5.1.5)), compute the first order perturbation theory correction to the eigenenergies of a harmonic oscillator perturbed by a potential $V(x) = g\delta(x)$.

5.2.6 *... and by a uniform field*

(a) *Find an exact expression for the first order perturbation theory correction to the ground state energy of a harmonic oscillator perturbed by a uniform field $V(x) = -Fx$.*

(b) Analyze the full Hamiltonian, including the perturbation. Observe that even with the perturbation, the Hamiltonian still describes a harmonic oscillator. *Compute the energy shift exactly and compare it with your result for the question (a).*

5.2.7 *Perturbative expansion of the expectation value of the perturbation itself and the virial theorem*

Consider again the Schrödinger equation for a particle in a "2q-tic" potential:

$$\hat{H} = \underbrace{-\frac{\hbar^2}{2m}\frac{\partial^2}{\partial x^2}}_{\hat{T}} + \underbrace{K_q x^{2q}}_{\hat{\mathcal{V}}}$$

$$q = 1,\ 2,\ 3,\ \ldots\ .$$

(a) *Using dimensional analysis, and without resorting to the WKB approximation, prove that the spectrum has the form*

$$E_n = \left(\frac{\hbar^2}{m}\right)^{\frac{q}{q+1}} K_q^{\frac{1}{q+1}} f(n)\,, \tag{5.22}$$

where $f(n)$ *is an unknown universal function;*

(b) The quantum virial theorem (9.8) introduced in Chapter 9 would predict that in our case the quantum expectation value of the potential energy in eigenstates is proportional to their energies, with a coefficient of proportionality of $1/(q+1)$:

$$\langle\psi_n|\hat{\mathcal{V}}|\psi_n\rangle = \frac{1}{q+1}E_n\,. \tag{5.23}$$

Let us now split, formally, the coupling constant K_q into a sum of a principal part and a "perturbative" part,

$$K_q = (K_q)_0 + \Delta K_q\,,$$

and reinterpret the Hamiltonian accordingly:

$$\hat{H} = \underbrace{-\frac{\hbar^2}{2m}\frac{\partial^2}{\partial x^2} + (K_q)_0 x^{2q}}_{\hat{H}_0} + \underbrace{\Delta K_q x^{2q}}_{\hat{V}}\,.$$

Let us now apply perturbation theory, treating the correction $\hat{V}$ as a perturbation. *Show that the law*

$$\langle\psi_n|\hat{V}|\psi_n\rangle^{(2)} = 2E_n^{(2)}\,,$$

proven in Problem 5.1.14, is consistent with the results (5.22) and (5.23). Here, $\Box^{(2)}$ stands for a second order perturbation theory prediction.

5.2.8   *A little theorem*

Consider a Hamiltonian that consists of two terms,

$$\hat{H} = \hat{S} + \alpha \hat{T} \,,$$

where the second term enters with a variable prefactor $\alpha > 0$, not necessarily small. Assume that its spectrum is bounded from below, and imagine that its ground state energy is proportional to α^{η}:

$$E_0 = \epsilon \times \alpha^{\eta} \,,$$

with $\epsilon > 0$. *Using the property that the second order perturbation theory shift of the ground state energy is always non-positive, prove that η is bounded as $0 \leq \eta \leq 1$.*

Remark: For example, for a harmonic oscillator, $\alpha = \omega^2$, and $\eta = 1/2$.

5.3   Background

5.3.1   *Matrix elements of operators in the WKB approximation*

One of the dozens of incarnations of the quantum-classical correspondence is the relationship between the quantum matrix elements and the classical Fourier components of observables.

The WKB approximation for the matrix elements of observables in one-dimensional systems reads:

$$|A_{n_1,n_2}| \approx |A^{(CM)}_{n_1-n_2}(\bar{E}_{n_1,n_2})| \,, \tag{5.24}$$

where

$$A^{(CM)}_m(E) \equiv \frac{1}{T(E)} \int_0^{T(E)} dt \, A(t \mid E) \exp[-im\omega(E)t] \tag{5.25}$$

is the mth Fourier component of the classical evolution of A at an energy E,

$$\bar{E}_{n_1,n_2} \equiv \frac{E_{n_1} + E_{n_2}}{2} \tag{5.26}$$

is the average between the energies of the states involved, and $T(E)$ and $\omega(E) = 2\pi/T(E)$ are the classical period and the classical frequency at the energy E, respectively.

Here, E_n is governed by the WKB rule:

$$\oint dx \, p(x; E_n) = 2\pi\hbar(n + \delta) \,, \tag{5.27}$$

where δ is determined by the boundary conditions.

For the diagonal matrix elements in particular, the phase of the matrix elements can be fixed as well:

$$A_{n,n} \approx \frac{1}{T(E)} \int_0^{T(E)} dt\, A(t \mid E) \,. \tag{5.28}$$

In Eq. 5.26, selecting the arithmetic mean of E_{n_1} and E_{n_2} as the choice for the classical energy $\bar{E}_{n_1,n_2}$ is completely arbitrary; any value in between the two energies will provide the same accuracy. Note, however, that the choice (5.26) (a) guarantees the "hermiticity" of the approximate matrix elements and (b) is probably the most manageable from a technical point of view.

In multi-dimensional quantum-ergodic systems, Eq. 5.28 becomes

$$A_{n,n} \approx \lim_{T\to\infty} \frac{1}{T} \int_0^T dt\, A(t \mid E) \,. \tag{5.29}$$

Furthermore, the across-the-spectrum variance of both off-diagonal and diagonal matrix elements there can be related to the classical autocorrelation function.[3]

The relationship (5.24) follows from the requirement that at short times, the time evolution of the quantum expectation values of observables must reproduce the classical evolution:

$$\langle A(t) \rangle = \sum_{n_1,n_2} \psi_{n_1}^\star \psi_{n_2} \langle n_1 | \hat{A} | n_2 \rangle \exp[i(E_{n_1} - E_{n_2})t/\hbar]$$

$$\approx \sum_m A_m^{(CM)}(E) \exp[+im\omega t]$$

To arrive at the expression (5.24) one (a) assumes that the initial state is represented by a smooth real non-negative wavepacket and (b) uses the semiclassical expression for the quantum energy differences $E_{n_1} - E_{n_2} \approx \omega(\bar{E}_{n_1,n_2})(n_1 - n_2)$.

The phases of the off-diagonal matrix elements in (5.24) remain undetermined. This is not a consequence of the deficiency of the method, but a result of an absence of convention. Recall that both the initial phase of the classical motion and the phases of the individual quantum states can be chosen at will. Once fixed, the ambiguity in (5.24) can be resolved.

As an example, let us try to derive a WKB expression for the matrix elements of the coordinate of a harmonic oscillator of frequency Ω and mass M and compare it with the exact result. The exact result reads

$$x_{n,n'} = \sqrt{\frac{\hbar}{2M\Omega}} \{ \sqrt{n}\,\delta_{n',n-1} + \sqrt{n+1}\,\delta_{n',n+1} \} \,. \tag{5.30}$$

---

[3]M. Feingold and A. Peres, Distribution of matrix elements of chaotic systems, *Physical Review A* **34**, 591 (1986).

Let us now look at the semiclassical recipe. The classical evolution of the coordinate reads

$$x(t|E) = x_0(E)\cos(\omega(E)t)\,,$$

where $x_0(E) = \sqrt{2E/(M\Omega^2)}$ is the amplitude of the oscillation and $\omega(E) = \Omega$ is the frequency of the oscillation. (Note that the absence of energy dependence of the frequency is a unique property of the harmonic oscillator). The Fourier transform (5.25) of $x(t|E)$ then gives

$$x_m = \frac{1}{2}x_0(E)(\delta_{m,+1} + \delta_{m,-1})\,.$$

Substituting this result into the quantum-classical correspondence rule (5.24) gives

$$|x_{n,n'}| \overset{n\to\infty}{\approx} \sqrt{\frac{\hbar}{2M\Omega}}\{\sqrt{n}\delta_{n',n-1} + \sqrt{n+1}\delta_{n',n+1}\}\,. \tag{5.31}$$

Notice that we managed to reproduce the exact result in its entirety. Note that while the correctness of the prediction for the leading and subleading orders in n was expected, the good agreement for the subsequent orders of the $1/n$ expansion is pure coincidence.

Among various observables, of particular interest is the probability density $\rho(x_0) \equiv |\psi(x_0)|^2$ at a particular point x_0; it can be formally regarded as the expectation value of the operator $\delta(x - x_0)$. The classical prediction for the probability density reads

$$\rho_{\mathrm{CM}}(x_0) = \frac{2}{T(E)v(x_0,\,E)}\,, \tag{5.32}$$

where $T(E) = 2\int_{x_{min}}^{x_{max}} dx/v(x,\,E)$ is the classical period, $v(x,\,E) = \sqrt{2(E - V(x))/m}$ is the classical velocity, and $V(x)$ is the potential energy. However the relationship between the classical and quantum densities is not as straightforward as in the case of smooth functions of coordinates, such as x, x^2, etc.; recall that $\delta(x - x_0)$ is a sharp function of x, and it may retain some sensitivity to the quantum oscillations of the density, even in the limit of high excitation numbers. Indeed, the proper correspondence is

$$\rho(x_0) = 2\rho_{\mathrm{CM}}(x_0)\cos(k(x_0,\,E)x_0 + \phi)\,, \tag{5.33}$$

where $k(x,\,E) \equiv \sqrt{2m(E - V(x))}/\hbar$ is the semiclassical wave vector, related to the de-Broglie wave length as $k = 2\pi/\lambda_{\mathrm{dB}}$, and E is the energy of the state $\psi(x)$; ϕ is a purely quantum phase that depends on the values of the potential in the whole classically allowed range, and on the boundary conditions. In the particular case of the left turning point a being a "soft wall" (a conventional turning point), $\phi = \int_a^{x_0} dx\, k(x,\,E) - \frac{\pi}{4}$.

5.3.2 *Perturbation theory: a brief summary*

Basic setting. Consider an unperturbed Schrödinger equation

$$\hat{H}_0|\psi_n^{(0)}\rangle = E_n^{(0)}|\psi_n^{(0)}\rangle \,.$$

and its perturbed version

$$\hat{H}|\psi_n\rangle = E_n|\psi_n\rangle \,,$$

where the Hamiltonian H consists mostly of $\hat{H}_0$, which can be treated exactly, and a small correction that renders exact solution impossible:

$$\hat{H} = \hat{H}_0 + \epsilon\hat{V} \,.$$

In what follows, both the eigenstates $|\psi_n^{(0)}\rangle$ and the eigenstates $|\psi_n\rangle$ are assumed to be normalized to unity:

$$\langle\psi_n^{(0)}|\psi_n^{(0)}\rangle = \langle\psi_n|\psi_n\rangle = 1$$

The perturbative expansion procedure. The procedure introduces, at the intermediate stages, an auxiliary unnormalized eigenstate $|\phi_n\rangle$ of the Hamiltonian $\hat{H}$: this state is proportional to the properly normalized eigenstate $|\psi_n\rangle$, its zeroth-order part *equals* the properly noramalized eigenstate of $\hat{H}^{(0)}$, and each subsequent order is orthogonal to the zeroth-order part:

$$E_n = E_n^{(0)} + \epsilon E_n^{(1)} + \epsilon^2 E_n^{(2)} + \cdots \qquad (5.34)$$

$$|\phi_n\rangle = \underbrace{|\phi_n^{(0)}\rangle}_{=|\psi_n^{(0)}\rangle} + \epsilon\underbrace{|\phi_n^{(1)}\rangle}_{\perp|\phi_n^{(0)}\rangle} + \epsilon^2\underbrace{|\phi_n^{(2)}\rangle}_{\perp|\phi_n^{(0)}\rangle} + \cdots$$

$$|\psi_n\rangle = A_n|\phi_n\rangle$$

$$A_n = \frac{1}{\sqrt{\langle\phi_n|\phi_n\rangle}} = A_n^{(0)} + \epsilon A_n^{(1)} + \epsilon^2 A_n^{(2)} + \cdots$$

$$|\psi_n\rangle = |\psi_n^{(0)}\rangle + \epsilon|\psi_n^{(1)}\rangle + \epsilon^2|\psi_n^{(2)}\rangle + \cdots$$

$$= A_n^{(0)}|\phi_n^{(0)}\rangle + \epsilon\left(A_n^{(0)}|\phi_n^{(1)}\rangle + A_n^{(1)}|\phi_n^{(0)}\rangle\right)$$

$$+ \epsilon^2\left(A_n^{(0)}|\phi_n^{(2)}\rangle + A_n^{(1)}|\phi_n^{(1)}\rangle + A_n^{(2)}|\phi_n^{(0)}\rangle\right) + \cdots \,.$$

Some results. The first and second order expressions for the energy and the first order for the eigenstate give

$$E_n^{(1)} = \langle\psi_n^{(0)}|\hat{V}|\psi_n^{(0)}\rangle \qquad (5.35)$$

$$|\psi_n^{(1)}\rangle = \sum_{n'\neq n}\frac{\langle\psi_{n'}^{(0)}|\hat{V}|\psi_n^{(0)}\rangle}{E_n^{(0)} - E_{n'}^{(0)}}|\psi_{n'}^{(0)}\rangle$$

$$E_n^{(2)} = \sum_{n'\neq n}\frac{|\langle\psi_{n'}^{(0)}|\hat{V}|\psi_n^{(0)}\rangle|^2}{E_n^{(0)} - E_{n'}^{(0)}} \,.$$

The zeroth, first and second orders for the normalization constant A_n, which appears at the intermediate stages of the derivation (5.34), read

$$A_n^{(0)} = 1 \tag{5.36}$$
$$A_n^{(1)} = 0$$
$$A_n^{(2)} = -\frac{1}{2} \sum_{n' \neq n} \frac{|\langle \psi_{n'}^{(0)} | \hat{V} | \psi_n^{(0)} \rangle|^2}{(E_n^{(0)} - E_{n'}^{(0)})^2} \, .$$

5.3.3 *Non-positivity of the second order perturbation theory shift of the ground state energy*

One particular property of the perturbation theory expansion for the eigenenergies of the Hamiltonian has been found to be particularly useful for discarding candidates for a ground state of a Hamiltonian, or even discarding a particular perturbation as being produced by a regular potential. Consider the expression for the second order perturbation theory shift, $E_n^{(2)}$, of energy (see (5.35)). Set $n = 0$, i.e. assume that the state $|\psi_n^{(0)}\rangle$ is the ground state of the unperturbed Hamiltonian:[4]

$$E_0^{(0)} < E_{n'>0}^{(0)} \, . \tag{5.37}$$

The sign of the individual terms in the expression for $E_n^{(2)}$ is fully determined by the sign of the their respective denominators: their respective numerators are strictly non-negative. For a ground state, all these denominators are negative. Thus,

> The second order perturbation theory shift of the ground state energy is always non-positive:
>
> $$E_0^{(2)} \leq 0 \, .$$

A typical application of the above property is presented in Problem 5.1.11: there, the non-positivity of the ground state energy allows one to unambiguously conclude that a particular boundary condition—relevant to fermionic atoms in cold waveguides—is not an approximation for any particular realistic potential.

---

[4]In the case of a degenerate ground state, degenerate perturbation theory (not considered here) will apply: it will automatically exclude cases where the inequality below becomes an equality.

Chapter 6

Variational Problems

6.1 Solved problems

6.1.1 *Inserting a wall*

Using variational reasoning, prove that adding an infinitely hight wall,

$$\hat{U} = \begin{cases} +\infty & x \leq x_0 \\ 0 & x > x_0 \end{cases},$$

to a regular potential $V(x)$ can only increase the ground state energy (see Fig. 6.1).

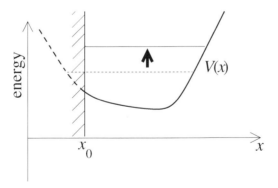

Fig. 6.1 Inserting a wall can only increase the ground state energy.

Solution: The ground state wavefunction, $\psi_0(x)$ minimizes the energy functional

$$\mathcal{E}[\psi(\cdot)] = \int dx \left\{ \frac{\hbar^2}{2m} |\psi'(x)|^2 + V(x)|\psi(x)|^2 \right\}$$

within a variational space of all normalized to unity continuous functions with piece-wise-continuous first derivative, $\mathfrak{V}$. In the presence of a wall, the ground state $\tilde{\psi}_0(x)$ minimizes the same functional within the space $\tilde{\mathfrak{V}}$ of all normalized to unity continuous functions with piece-wise-continuous first derivative that are equal to zero for $x \leq x_0$. Since $\tilde{\mathfrak{V}}$ is a subset of $\mathfrak{V}$, then the minimum of $E[\psi(\cdot)]$ among the members of $\tilde{\mathfrak{V}}$ can never be lower than the minimum of $E[\psi(\cdot)]$ among the members of $\mathfrak{V}$.

$$\boxed{Q.E.D.}$$

6.1.2   *Parity of the eigenstates*

Can the ground state $\psi_0(x)$ of the sextic potential $V(x) = \alpha x^6$ be odd, i.e. can it be that $\psi_0(-x) = -\psi_0(x)$? Explain your answer.
    Solution:

$$\boxed{\text{The ground state has no nodes} \Rightarrow \text{it can not be odd}}$$

(See Sec. 6.1.5 below.)

6.1.3   *Simple variational estimate for the ground state energy of a harmonic oscillator*

Give a variational estimate for the ground state energy of a harmonic oscillator,

$$\hat{H} = -\frac{\hbar^2}{2m}\frac{\partial^2}{\partial x^2} + \frac{m\omega^2}{2}x^2 ,$$

using the following variational ansatz:

$$\psi_a(x) = \sqrt{\frac{3}{2a}} \times \begin{cases} 1 - \frac{|x|}{a} , & |x| \leq a \\ 0 , & |x| > a \end{cases} . \tag{6.1}$$

    Solution: The ground state constitutes a point of minimum of the variational functional

$$\mathcal{E}[\psi(\cdot)] = \int dx \left\{ \frac{\hbar^2}{2m}(\psi'(x))^2 + \frac{m\omega^2}{2}x^2(\psi(x))^2 \right\}$$

over the space of all continuous, piecewise differentiable real functions normalized to unity:

$$\int_{-\infty}^{+\infty} dx (\psi(x))^2 = 1 .$$

Observe that the ansatz (6.1) does belong to this space. The minimum value of the functional equals the ground state energy.

Over the functions in the one-dimensional variational space (6.1), the energy functional (6.2) attains a value

$$\mathcal{E}[\psi_a(\cdot)] = \frac{3}{2}\frac{\hbar^2}{ma^2} + \frac{1}{20}m\omega^2 a^2$$

Finally, minimizing this energy over possible values of the width a, we get

$$\mathcal{E}_{\min} = \sqrt{\frac{3}{10}}\,\hbar\omega = 0.547\ldots \times \hbar\omega$$

The variational ground state energy is only 10% higher than the exact value of $0.5\hbar\omega$. This is a good result given how simple the variational ansatz was.

6.1.4 *A property of variational estimates*

Consider the motion of a single one-dimensional particle in a generic potential well $V(x)$ (see Fig. 6.2). Consider two variational estimates for the ground state energy, $E_{0,\,1\text{-param.}}$ and $E_{0,\,2\text{-param.}}$ obtained using the ansatz

$$\psi_{0,\,1\text{-param.}}(x) = \text{const} \times \begin{cases} 1 - \dfrac{x}{a}\,, & |x| < a \\[2mm] 0 & ,\ |x| > a \end{cases}$$

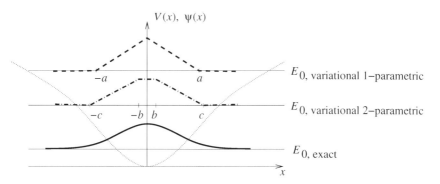

Fig. 6.2   One-parametric and two-parametric variational piecewise linear anzatses for a generic symmetric one-dimensional potential.

and the ansatz

$$\psi_{0,\,2\text{-param.}}(x) = \text{const} \times \begin{cases} 1 - \dfrac{b}{c}\,, & |x| < b \\[2mm] 1 - \dfrac{x}{c}\,, & b < |x| < c \\[2mm] 0 & , \quad |x| > c \end{cases}$$

respectively.

Prove that the one-parametric estimate can never be lower than the two-parametric one:

$$E_{0,\,1\text{-param.}} \geq E_{0,\,2\text{-param.}}$$

Hint: *Go through our derivation of the fact that the true ground state energy is less than (or equal to) any of its variational approximations and try to mimic it.*

Solution: Figure 6.3 illustrates the proof. The line of reasoning goes as follows:

- The 2-parametric variational space $\mathfrak{V}^{(2)}$ includes the 1-parametric one, $\mathfrak{V}^{(1)}$:

$$\mathfrak{V}^{(1)} \subset \mathfrak{V}^{(2)}$$

- Thus, the 1-parametric variational approximation (for the ground state), $|\psi_{\text{g.s.}}^{(1)}\rangle$ belongs to $\mathfrak{V}^{(2)}$ as well:

$$|\psi_{\text{g.s.}}^{(1)}\rangle \in \mathfrak{V}^{(2)}$$

- By construction, the energy of the 2-parametric variational approximation is less than or equal to the energy of any state in $\mathfrak{V}^{(2)}$:

$$E[|\psi_{\text{g.s.}}^{(2)}\rangle] \leq E[|\psi\rangle \in \mathfrak{V}_2]$$

- In particular, this should hold for $|\psi_{\text{g.s.}}^{(1)}\rangle$:

$$E[|\psi_{\text{g.s.}}^{(2)}\rangle] \leq E[|\psi_{\text{g.s.}}^{(1)}\rangle]$$

- $\boxed{Q.E.D.}$

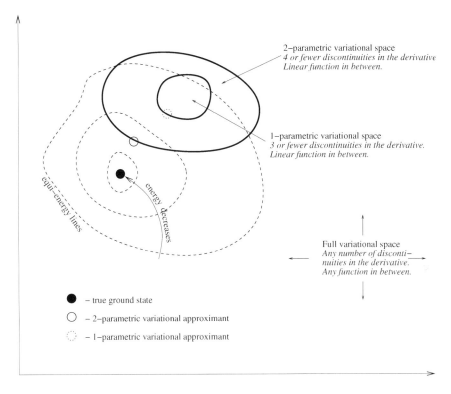

Fig. 6.3   An illustration to the solution to Problem 6.1.4.

6.1.5   *Absence of nodes in the ground state*

Consider a Hamiltonian

$$\hat{H} = -\frac{\hbar^2}{2m}\frac{\partial^2}{\partial x^2} + V(x).$$

Prove variationally that its ground state wavefunction $\Psi_0(x)$ *does not have nodes.*

<u>Solution:</u> Both real and imaginary parts of the ground state wavefunction $\Psi_0(x)$ are minimal points of an energy functional

$$\mathcal{E}_{r2}[\psi(\cdot)] \equiv \frac{\frac{\hbar^2}{2m}\int_{-\infty}^{+\infty} dx\,(\psi'(x))^2 + \int_{-\infty}^{+\infty} dx\, V(x)\psi^2(x)}{\int_{-\infty}^{+\infty} dx\,\psi^2(x)}$$

acting on real wavefunctions. In what follows $\psi_0(x)$ stands for either real or imaginary part of the wavefunction $\Psi_0(x)$, and we will prove that neither of them can have nodes.

Proof by reductio ad absurdum. To prove the absence of nodes, we are going to show that if a node is present, then it is possible to construct a state $\tilde{\psi}(x)$ whose energy, $\mathcal{E}_{r2}[\tilde{\psi}(\cdot)]$, is lower than the ground state energy $\mathcal{E}_{r2}[\psi_0(\cdot)]$.

The construction of the state $\tilde{\psi}(x)$ proceeds as follows. Assume that $\psi_0(x)$ has a node at a point x_0. Consider an intermediate state

$$\tilde{\tilde{\psi}}(x) = \begin{cases} -\psi_0(x) & \text{for } x \leq x_0 \\ +\psi_0(x) & \text{for } x \geq x_0 \end{cases}$$

(see Fig. 6.4). The energy of this state is the same as the energy of $\psi_0(x)$:

$$\mathcal{E}_{r2}[\tilde{\tilde{\psi}}(\cdot)] = \mathcal{E}_{r2}[\psi_0(\cdot)]. \tag{6.2}$$

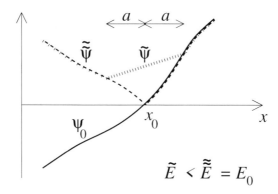

Fig. 6.4   An illustration to the proof of the absence of nodes in the ground state. (See explanations in text.)

Next, we construct a state

$$\tilde{\psi}(x) = \begin{cases} \frac{1}{2}(\tilde{\tilde{\psi}}(x_0 + a) + \tilde{\tilde{\psi}}(x_0 - a)) + \frac{x - x_d}{2a}(\tilde{\tilde{\psi}}(x_0 + a) - \tilde{\tilde{\psi}}(x_0 - a)) \\ \qquad\qquad\qquad\qquad\qquad \text{for } x_0 - a \leq x \leq x_0 + a \\ \\ \tilde{\tilde{\psi}}(x) , \quad \text{otherwise,} \end{cases},$$

for some interval of size a. For a sufficiently small a, the energy of the state $\tilde{\psi}(x)$,

$$\mathcal{E}_{r2}[\tilde{\psi}(\cdot)] = \mathcal{E}_{r2}[\tilde{\tilde{\psi}}(\cdot)] - \frac{\hbar^2}{m}(\psi_0'(x_0))^2 a + \mathcal{O}(a^3),$$

becomes lower than the energy of $\tilde{\tilde{\psi}}(x)$:

$$\mathcal{E}_{r2}[\tilde{\psi}(\cdot)] < \mathcal{E}_{r2}[\tilde{\tilde{\psi}}(\cdot)]. \tag{6.3}$$

Combining the relationships (6.2) and (6.3) we arrive at a contradiction to the assumption that $\psi_0(x)$ minimizes the energy:

$$\mathcal{E}_{r2}[\tilde{\psi}(\cdot)] < \mathcal{E}_{r2}[\psi_0(\cdot)].$$

Recall that $\psi_0(x)$ stands for either real or imaginary part of the ground state wavefunction $\Psi_0(x)$.

$$\boxed{Q.E.D.}$$

Remark: The discontinuity in the derivative of $\tilde{\tilde{\psi}}(x)$ also contradicts a more general statement on the continuity of derivatives in states minimizing the energy functional $\mathcal{E}_{r2}$ (see Problem 6.4.2).

6.1.6 *Absence of degeneracy of the ground state energy level*

Consider a Hamiltonian

$$\hat{H} = -\frac{\hbar^2}{2m}\frac{\partial^2}{\partial x^2} + V(x).$$

Using the conclusion of Problem 6.1.5, prove that the ground state of this Hamiltonian is not degenerate.

Solution: Proof by reductio ad absurdum. Assume that there are two distinct eigenstates, $\Psi_{0,I}(x)$ and $\Psi_{0,II}(x)$, that correspond to the ground state energy E_0. Let us construct another ground state,

$$\tilde{\Psi}_0(x) \equiv \Psi_{0,II}(x_0)\Psi_{0,I}(x) - \Psi_{0,I}(x_0)\Psi_{0,II}(x),$$

for some spatial point x_0. This state will have a node at $x = x_0$. Thus, according to the statement proven in Problem 6.1.5, this state can not be a ground state, and we arrive at a contradiction.

$$\boxed{Q.E.D.}$$

6.2 Problems without provided solutions

6.2.1 *Do stronger potentials always lead to higher ground state energies?*

Prove variationally that if

$$V_2(x) \geq V_1(x)$$

everywhere, then the ground state energy of a Hamiltonian $\hat{H}_2$ is always greater than or equal to the one for a Hamiltonian $\hat{H}_1$:

$$E_{\text{g.s., 2}} \geq E_{\text{g.s., 1}}.$$

Here

$$\hat{H}_\alpha \equiv -\frac{1}{2}\frac{\partial^2}{\partial x^2} + V_\alpha(x).$$

6.2.2 *Variational analysis meets perturbation theory*

Prove variationally that the first order perturbation theory correction to the ground state energy is given by the expectation value of the perturbation in an unperturbed ground state:

$$E_{\text{g.s.}}^{(1)} = \langle \psi_{\text{g.s.}}^{(0)} | \hat{V} | \psi_{\text{g.s.}}^{(0)} \rangle.$$

6.2.3 *Another variational estimate for the ground state energy of a harmonic oscillator ...*

Give a variational estimate for the ground state energy of a harmonic oscillator,

$$\hat{H} = -\frac{\hbar^2}{2m}\frac{\partial^2}{\partial x^2} + \frac{m\omega^2}{2}x^2,$$

using the following variational ansatz:

$$\psi_a(x) = \text{const} \times e^{-|x|/a}.$$

Do not forget to normalize it first.

6.2.4 *... and yet another*

Now use the Gaussian variational anzats:

$$\psi_a(x) = \text{const} \times e^{-x^2/a^2}.$$

Show that it predicts the *exact* value for the ground state energy.

6.2.5 Gaussian- and wedge- variational ground state energy of a quartic oscillator

(a) *Show that the Gaussian variational prediction for the ground state energy of a quartic oscillator,*

$$-\frac{\hbar^2}{2m}\frac{\partial^2}{\partial x^2}\psi(x) + \beta x^4 \psi(x) = E\psi(x),$$

is

$$E_{g.s.} = \frac{3 \times 3^{1/3}}{4 \times 2^{2/3}}\left(\frac{\beta\hbar^4}{m^2}\right)^{1/3}.$$

(b) *Now, use the "wedge" ansatz (6.1) to estimate $E_{g.s.}$.*

(c) *Determine which of the two answers, (a) or (b), is closer to the exact ground state energy.*

6.3 Background

6.3.1 Variational analysis

The ground state $|\Psi_0\rangle$ of a Hamiltonian $\hat{H}$ constitutes a *global minimum point* of a *variational functional*

$$\mathcal{E}[|\Psi\rangle] \equiv \frac{\langle\Psi|\hat{H}|\Psi\rangle}{\langle\Psi|\Psi\rangle}, \tag{6.4}$$

over a *variational space* formed by all members of the Hilbert space. The ground state energy E_0, as in $\hat{H}|\Psi\rangle = E_0|\Psi\rangle$, constitutes a global minimum of the functional (6.4).

For a one-dimensional Hamiltonian

$$\hat{H} = -\frac{\hbar^2}{2m}\frac{\partial^2}{\partial x^2} + V(x); \tag{6.5}$$

the above roles are distributed as follows:

Variational functional:

$$\mathcal{E}[\Psi(\cdot)] \equiv \frac{-\frac{\hbar^2}{2m}\int_{-\infty}^{+\infty}dx\,\Psi^\star(x)\Psi''(x) + \int_{-\infty}^{+\infty}dx\,|\Psi|^2(x)V(x)}{\int_{-\infty}^{+\infty}dx\,|\Psi|^2(x)}; \tag{6.6}$$

Variational space $\mathfrak{V}$: space of all complex-valued functions $\Psi(x)$ with continuous first derivative.

Problem 6.4.1 below illustrates that the real and imaginary parts of each[1] of the global minimum points Ψ_0 corresponding to a global minimum E_0 of the functional $\mathcal{E}[\Psi(\cdot)]$ (6.6) are global minima of a

Variational functional

$$\mathcal{E}_r[\psi(\cdot)] \equiv \frac{-\frac{\hbar^2}{2m} \int_{-\infty}^{+\infty} dx\, \psi(x)\psi''(x) + \int_{-\infty}^{+\infty} dx\, V(x)\psi^2(x)}{\int_{-\infty}^{+\infty} dx\, \psi^2(x)},$$

(6.7)

over a

Variational space $\mathfrak{V}_r$: space of all real-valued functions $\psi(x)$ with continuous first derivative.

At both the real and imaginary parts of Ψ_0, the real functional (6.7) is equal to E_0.

Within the variational space $\mathfrak{V}_r$, the values of the functional (6.7) coincide with the ones for the . . .

Variational functional

$$\mathcal{E}_{r2}[\psi(\cdot)] \equiv \frac{\frac{\hbar^2}{2m} \int_{-\infty}^{+\infty} dx\, (\psi'(x))^2 + \int_{-\infty}^{+\infty} dx\, V(x)\psi^2(x)}{\int_{-\infty}^{+\infty} dx\, \psi^2(x)},$$

(6.8)

provided the wavefunction decays to zero at $\pm\infty$. It is tempting to extend the variational space to a broader

Variational space $\mathfrak{V}_{r2}$: space of all real-valued continuous functions $\psi(x)$.

The $\mathfrak{V}_{r2}$ extension has at least two advantages over the standard space $\mathfrak{V}_r$: (a) the technically more manageable piecewise linear continuous functions can be used as variational ansatzes; (b) several general theorems can be proved using the space $\mathfrak{V}_{r2}$ (see Problems 6.1.5 and 6.1.6).

---

[1]It can be shown variationally (see for example Problem 6.1.6) that in reality, the ground state of the Hamiltonian (6.5) is never degenerate, and thus, the minima of the functional (6.6) are different from each other by a trivial multiplicative factor. However, the result proven in Problem 6.4.1 can be extended to any real Hamiltonian. On the other hand, the absence of degeneracy is a property specific to Hamiltonians with a kinetic energy density $\tau(\psi'(x))$ that have a non-negative second derivative $d^2\tau(\xi)/d\xi^2$ for all real ξ.

It is yet to be proven, however, that the global minimum point of $\mathcal{E}_r[\psi(\cdot)]$ is also a global minimum point of the extended functional $\mathcal{E}_{r2}[\psi(\cdot)]$. Otherwise, it may so happen that a particular ansatz belongs to a well around an *unphysical* minimum, and thus predicts an energy that has nothing to do with the "true" ground state energy. Problem 6.4.2 is devoted to a proof that all minimum points of $\mathcal{E}_{r2}[\psi(\cdot)]$ have continuous first derivatives, and thus the functional $\mathcal{E}_{r2}[\psi(\cdot)]$ can indeed be used to identify the minima of the functional $\mathcal{E}_r[\psi(\cdot)]$.

6.4 Problems linked to the "Background"

6.4.1 *Complex vs. real variational spaces*

Prove that the real and imaginary parts of each of the global minimum points Ψ_0 corresponding to a global minimum E_0 of the functional $\mathcal{E}[\Psi(\cdot)]$ (6.6) are global minimum points of the functional $\mathcal{E}_r[\psi(\cdot)]$ (6.7); at these points, the functional $\mathcal{E}_r$ equals E_0.

   <u>Solution:</u> (1) Represent $\Psi_0(x)$ as $\Psi_0(x) = u_0(x) + iv_0(x)$ ($u_0(x) \equiv \Re(\Psi_0(x))$, $v_0(x) \equiv \Im(\Psi_0(x))$).

(2) Observe that for functions $\Psi(x)$ that decay to zero as $x \to \pm\infty$, the functionals $\mathcal{E}[\Psi(\cdot)]$ and $\mathcal{E}_r[\psi(\cdot)]$ can be represented as

$$\mathcal{E}[\Psi(\cdot)] \equiv \frac{\frac{\hbar^2}{2m}\int_{-\infty}^{+\infty}dx\,|\Psi'(x)|^2 + \int_{-\infty}^{+\infty}dx\,V(x)|\Psi|^2(x)}{\int_{-\infty}^{+\infty}dx\,|\Psi|^2(x)},$$

$$\mathcal{E}_r[\psi(\cdot)] \equiv \frac{\frac{\hbar^2}{2m}\int_{-\infty}^{+\infty}dx\,(\psi'(x))^2 + \int_{-\infty}^{+\infty}dx\,V(x)\psi^2(x)}{\int_{-\infty}^{+\infty}dx\,\psi^2(x)}.$$

This can easily be proven using integration by parts.

(3) Since Ψ_0 is a global minimum point, then

$$\mathcal{E}[\Psi(\cdot)] \geq \mathcal{E}[\Psi_0(\cdot)]$$

for any state $\Psi(x)$.

(4) Consider first $\Psi(x)$ of a form

$$\Psi(x) = \tilde{u}(x) + i\,v_0(x),$$

where $\tilde{u}(x)$ is any real function normalized the same way as $u_0(x)$:

$$\int_{-\infty}^{+\infty} dx\,\tilde{u}^2(x) = \int_{-\infty}^{+\infty} dx\,u_0^2(x).$$

(5) Then the following chain applies:

$$\frac{\frac{\hbar^2}{2m}\int_{-\infty}^{+\infty}dx\,\left((\tilde{u}'(x))^2 + (v_0'(x))^2\right) + \int_{-\infty}^{+\infty}dx\,V(x)\left(\tilde{u}^2(x) + v_0^2(x)\right)}{\int_{-\infty}^{+\infty}dx\,\left(\tilde{u}^2(x) + v_0^2(x)\right)}$$

$$\overset{(3)}{\geq} \frac{\frac{\hbar^2}{2m}\int_{-\infty}^{+\infty}dx\,\left((u_0'(x))^2 + (v_0'(x))^2\right) + \int_{-\infty}^{+\infty}dx\,V(x)\left(u_0^2(x) + v_0^2(x)\right)}{\int_{-\infty}^{+\infty}dx\,\left(u_0^2(x) + v_0^2(x)\right)}$$

$$\Downarrow\leftarrow (4)$$

$$\frac{\hbar^2}{2m}\int_{-\infty}^{+\infty}dx\,(\tilde{u}'(x))^2 + \int_{-\infty}^{+\infty}dx\,V(x)\tilde{u}^2(x)$$

$$\geq \frac{\hbar^2}{2m}\int_{-\infty}^{+\infty}dx\,(u_0'(x))^2 + \int_{-\infty}^{+\infty}dx\,V(x)u_0^2(x)$$

$$\Downarrow$$

$$\mathcal{E}_r[\tilde{u}(\cdot)] \geq \mathcal{E}_r[u_0(\cdot)]\,.$$

(6) Let us now prove that

$$\mathcal{E}_r[u_0(\cdot)] = \mathcal{E}[\Psi_0(\cdot)]\,.$$

This can be proved by demonstrating that $u_0(x)$ is a global minimum of $\mathcal{E}$. The proof goes as follows:

$$\mathcal{E}[\Psi(\cdot)]$$

$$\overset{\Psi\equiv u+iv}{=} \frac{\frac{\hbar^2}{2m}\int_{-\infty}^{+\infty}dx\,\left((u'(x))^2 + (v'(x))^2\right) + \int_{-\infty}^{+\infty}dx\,V(x)\left(u^2(x) + v^2(x)\right)}{\int_{-\infty}^{+\infty}dx\,\left(u^2(x) + v^2(x)\right)}$$

$$= \frac{\mathcal{E}_r[u(\cdot)]\int_{-\infty}^{+\infty}dx\,u^2(x) + \mathcal{E}_r[v(\cdot)]\int_{-\infty}^{+\infty}dx\,v^2(x)}{\int_{-\infty}^{+\infty}dx\,u^2(x) + \int_{-\infty}^{+\infty}dx\,v^2(x)}$$

$$\overset{\mathcal{E}_r[u(\cdot)]\geq\mathcal{E}_r[u_0(\cdot)]}{\geq} \frac{\mathcal{E}_r[u_0(\cdot)]\int_{-\infty}^{+\infty}dx\,u^2(x) + \mathcal{E}_r[v(\cdot)]\int_{-\infty}^{+\infty}dx\,v^2(x)}{\int_{-\infty}^{+\infty}dx\,u^2(x) + \int_{-\infty}^{+\infty}dx\,v^2(x)}$$

$$\overset{\mathcal{E}_r[v(\cdot)]\geq\mathcal{E}_r[u_0(\cdot)]}{\geq} \frac{\mathcal{E}_r[u_0(\cdot)]\int_{-\infty}^{+\infty}dx\,u^2(x) + \mathcal{E}_r[u_0(\cdot)]\int_{-\infty}^{+\infty}dx\,v^2(x)}{\int_{-\infty}^{+\infty}dx\,u^2(x) + \int_{-\infty}^{+\infty}dx\,v^2(x)}$$

$$= \mathcal{E}_r[u_0(\cdot)]\,.$$

(7) For an arbitrary real part $u(x)$ of the wavefunction $\Psi(x)$, one may consider its normalized counterpart

$$\tilde{u}(x) \equiv \sqrt{\frac{\int_{-\infty}^{+\infty}dx\,u_0^2(x)}{\int_{-\infty}^{+\infty}dx\,u^2(x)}}\,u(x)\,.$$

The value of $\mathcal{E}_r$ at $u(x)$ is the same as at $\tilde{u}(x)$. Thus

$$\mathcal{E}_r[u(\cdot)] = \mathcal{E}_r[\tilde{u}(\cdot)] \geq \mathcal{E}_r[u_0(\cdot)]$$
$$\forall u(x).$$

For the imaginary part, the proof is completely analogous.

6.4.2 *A proof that the $(\psi')^2$ energy functional does not have minima with discontinuous derivatives*

Prove that the minimum points of the functional $\mathcal{E}_{r2}$ (6.8) correspond to functions with continuous first derivatives.

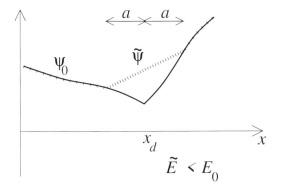

Fig. 6.5 An illustration to the proof of the absence of discontinuities of derivative in a local minimum of the energy functional $\mathcal{E}_{r2}[\psi(\cdot)]$. (See explanations in text.)

Solution: Proof by reductio ad absurdum. The idea of the proof is to show that for would-be minima with discontinuous derivatives, one can always find a small deviation that leads to a further decrease in the value of the energy functional.

Assume that the function $\psi_0(x)$ is a local minimum of the functional

$$\mathcal{E}_{r2}[\psi(\cdot)] \equiv \frac{\frac{\hbar^2}{2m} \int_{-\infty}^{+\infty} dx \, (\psi'(x))^2 + \int_{-\infty}^{+\infty} dx \, V(x)\psi^2(x)}{\int_{-\infty}^{+\infty} dx \, \psi^2(x)} ,$$

and that it has a discontinuous derivative at a point x_d. Assume, without loss of generality, that ψ_0 is normalized to unity:

$$\int_{-\infty}^{+\infty} dx \, \psi_0^2(x) = 1 .$$

Consider the following state:

$$\tilde{\psi}(x) = \begin{cases} \frac{1}{2}(\psi_0(x_d + a) + \psi_0(x_d - a)) + \frac{x - x_d}{2a}(\psi_0(x_d + a) - \psi_0(x_d - a)), \\ \qquad\qquad\qquad\qquad\qquad \text{for } x_d - a \le x \le x_d + a \\[2ex] \psi_0(x), \quad \text{otherwise,} \end{cases}$$

where $a \ge 0$ (see Fig. 6.5). The value of the energy functional at $\psi(x)$ is

$$\begin{aligned} \mathcal{E}_{r2}[\tilde{\psi}(\cdot)] &= \mathcal{E}_{r2}[\psi_0(\cdot)] - \frac{\hbar^2}{4m}(\psi_0'(x_d+) - \psi_0'(x_d-))^2 a \\ &\quad + (\psi_0'(x_d+) - \psi_0'(x_d-)) \\ &\qquad \times \left(-\frac{\hbar^2}{2m}(\psi_0''(x_d+) + \psi_0''(x_d-)) + V(x_d) - \mathcal{E}_{r2}[\psi_0(\cdot)] \right) a^2 + \mathcal{O}(a^3). \end{aligned}$$

$$(6.9)$$

Observe that the prefactor in front of a is strictly non-positive, and it reaches zero *only* if the first derivative were continuous at x_d. Contrary to the assumption that the energy functional $\mathcal{E}_{r2}$ reaches a minimum at $\psi_0(x)$, one can always find a small but finite interval of size a such that the value of $\mathcal{E}_{r2}$ at $\tilde{\psi}(x)$ is lower than that at $\psi_0(x)$. The only way to resolve this contradiction is for the first derivative to be continuous.

$$\boxed{Q.E.D.}$$

Remark 1: Observe an interesting trend. The function $\tilde{\psi}(x)$ can be regarded as an attempt to "smooth" the discontinuity in the derivative. The non-positivity of the leading term in the expansion (6.9) indicates that "curing" the discontinuities will allow a lower energy, and that the global minimum of energy is attained at a state with continuous first derivative.

Remark 2: It can be shown that in the case of continuous first derivative, the deviation between $\mathcal{E}_{r2}[\tilde{\psi}(\cdot)]$ and $\mathcal{E}_{r2}[\psi_0(\cdot)]$ starts as late as order of a^3.

Chapter 7

Gravitational Well: a Case Study

Introduction

Consider a Schrödinger equation for a particle of mass m jumping on a floor in a gravitational field (Fig. 7.1):

$$-\frac{\hbar^2}{2m}\frac{\partial^2}{\partial x^2}\psi(x) + U(x)\psi(x) = E\psi(x)$$

$$U(x) = \begin{cases} +\infty & \text{for } x < 0 \\ \alpha\,x & \text{for } x \geq 0 \end{cases}, \qquad (7.1)$$

where $\alpha = mg$ is the coupling constant, g is the local gravitational constant, and x is the elevation. In what follows, we will study the spectrum of the system and the response of this spectrum to a small change in the coupling constant. The exact diagonalization of the problem is relatively complicated: in particular, it involves the Airy function and its zeros. Instead of the exact values, we will be looking for approximate values of the observables in question; to this end, we will be employing both the WKB approximation and perturbation theory.

7.1 Solved problems

7.1.1 Bohr-Sommerfeld quantization

Using the Bohr-Sommerfeld quantization rule, determine the spectrum of the system, E_n.

Solution: There is one hard and one soft turning point. Thus

$$\oint p(x|E_n)\,dx = 2\pi\hbar(n - 1/4) \qquad (7.2)$$

$$n = 1,\,2,\,3,\,\ldots\,.$$

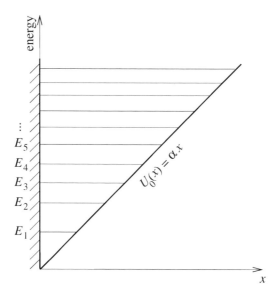

Fig. 7.1   A gravitational well.

The l.h.s. integral in (7.2) gives

$$\oint p\,dx = 2\int_0^{E_n/\alpha}\sqrt{2m(E_n - \alpha x)}$$
$$= \frac{4\sqrt{2}}{3}\frac{\sqrt{m}}{\alpha}E_n^{3/2}\,.$$

We get finally

$$E_n = \frac{(3\pi)^{2/3}}{2}\left(\frac{\hbar^2\alpha^2}{m}\right)^{1/3}(n - 1/4)^{2/3} \qquad (7.3)$$

7.1.2   *A WKB-based order-of-magnitude estimate for the spectrum*

By replacing the phase-space trajectory with a rectangle, estimate the spectrum.

<u>Solution:</u> The l.h.s. integral in (7.2) gives

$$\oint p \, dx \sim 2 \Delta p \, \Delta x$$

$$\sim \frac{\sqrt{m}}{\alpha} E_n^{3/2} \, ,$$

where $\Delta p \sim \sqrt{m E_n}$ and $\Delta x \sim E_n/\alpha$, both estimates being derived from energy conservation.

We get finally

$$\boxed{E_n \sim \left(\frac{\hbar^2 \alpha^2}{m} \right)^{1/3} (n - 1/4)^{2/3}}$$

7.1.3 A WKB-based dimensional estimate for the spectrum

Starting from the Bohr-Sommerfeld rule, use dimensional analysis to estimate the spectrum.

<u>Solution:</u> Introduce $\tilde{\eta} \equiv (\hbar(n - 1/4))^2/2m$. $\tilde{\eta}$ and α are the only two input parameters entering the quantization rule (7.2). The unknown is E_n.

– *The principal units—the units of length and the units of energy:*

$$[\mathcal{L}], \quad [\mathcal{E}];$$

– *The input parameters and their units:*

$$[\tilde{\eta}] = [\mathcal{L}]^2 [\mathcal{E}]$$
$$[\alpha] = [(1/\mathcal{L})] [\mathcal{E}] \, ;$$

– *The set of independent dimensionless parameters* $= \emptyset$;
– *The principal scales—the length scale and the energy scale, examples of:*

$$\mathcal{L} = \left(\frac{\hbar^2}{m\alpha} \right)^{1/3}$$

$$\mathcal{E} = \left(\frac{\hbar^2 \alpha^2}{m} \right)^{1/3};$$

– *Solution for the unknown:*

$$[E_n] = [\mathcal{E}] \Rightarrow E_n \sim \mathcal{E} = \left(\frac{\hbar^2 \alpha^2}{m} \right)^{1/3}.$$

Thus,

$$E_n \sim \left(\frac{\hbar^2 \alpha^2}{m}\right)^{1/3} (n - 1/4)^{2/3}$$

7.1.4   *A perturbative calculation of the shift of the energy levels under a small change in the coupling constant. The first order*

Split the coupling constant α as

$$\alpha = \alpha_0 + \Delta\alpha \ ,$$

and reinterpret the problem (7.1) as

$$\underbrace{\left(-\frac{\hbar^2}{2m}\frac{\partial^2}{\partial x^2} + U_0(x)\right)}_{\text{principal part, } \hat{H}_0} \psi(x) + \underbrace{V(x)}_{\text{perturbation, } \hat{V}} \psi(x) = E\psi(x), \quad (7.4)$$

where

$$U_0(x) = \begin{cases} +\infty & \text{for } x < 0 \\ \alpha_0\, x & \text{for } x \geq 0 \end{cases},$$

$$V(x) = \begin{cases} +\infty & \text{for } x < 0 \\ \Delta\alpha\, x & \text{for } x \geq 0 \end{cases}, \quad (7.5)$$

and we assume that $\Delta\alpha$ is small: $\Delta\alpha \ll \alpha_0$. Its unperturbed spectrum obviously reads

$$E_n^{(0)} = \frac{(3\pi)^{2/3}}{2}\left(\frac{\hbar^2 \alpha_0^2}{m}\right)^{1/3}(n - 1/4)^{2/3}.$$

Using perturbation theory, calculate the first order correction to the spectrum, $E_n^{(1)}$. Regard $V(x)$ as a small perturbation. Use the WKB expressions for the matrix elements of the relevant observables.

Solution: The first order of perturbation theory expansion for the energy reads

$$\begin{aligned} E_n^{(1)} &= \langle \psi_n^{(0)} | \hat{V} | \psi_n^{(0)} \rangle \\ &= \Delta\alpha \, \psi_n^{(0)} |x| \psi_n^{(0)} \\ &\approx \frac{1}{T(E_n^{(0)})} \oint_{T(E_n^{(0)})} dt\, x(t|E_n^{(0)}), \end{aligned}$$

where, following the quantum-classical correspondence, the (quantum) diagonal matrix element of the coordinate x is substituted by its classical temporal average. Here, the integral $\oint$ covers one full cycle of motion. The classical trajectory $x(t|E)$ is taken at the unperturbed energy $E_n^{(0)}$, calculated in turn using the WKB approximation.

The classical average gives,

$$\frac{1}{T(E_n^{(0)})} \oint_{T(E_n^{(0)})} dt\, x(t|E_n^{(0)})$$

$$= \underbrace{\left(\frac{2mv_0(E_n^{(0)})}{\alpha_0}\right)^{-1}}_{1/T(E_n^{(0)})} \int_0^{\frac{2mv_0(E_n^{(0)})}{\alpha_0}} dt \underbrace{\left(v_0(E_n^{(0)})t - \frac{1}{2}\frac{\alpha_0}{m}t^2\right)}_{x(t|E_n^{(0)})}$$

$$= \frac{2}{3}\left(\frac{\Delta\alpha}{\alpha_0}\right)E_n^{(0)},$$

where $v_0(E) = \sqrt{2E/m}$ is the initial velocity. Notice that this result is consistent with the previously derived scaling of the spectrum E_n with the coupling constant α, $E_n \propto \alpha^{2/3}$. Indeed, replacing α with $\alpha_0 + \Delta\alpha$, we get

$$E_n\big|_{\alpha_0 + \Delta\alpha} = \left(\frac{\alpha_0 + \Delta\alpha}{\alpha_0}\right)^{2/3} E_n\big|_{\alpha_0}$$

$$= \left(1 + \frac{2}{3}\left(\frac{\Delta\alpha}{\alpha_0}\right) - \frac{2}{9}\left(\frac{\Delta\alpha}{\alpha_0}\right)^2 + \cdots\right)E_n\big|_{\alpha_0}$$

$$= \underbrace{E_n\big|_{\alpha_0}}_{E_n^{(0)}} + \underbrace{\frac{2}{3}\left(\frac{\Delta\alpha}{\alpha_0}\right)E_n\big|_{\alpha_0}}_{E_n^{(1)}} - \underbrace{\frac{2}{9}\left(\frac{\Delta\alpha}{\alpha_0}\right)^2 E_n\big|_{\alpha_0}}_{E_n^{(2)}} + \cdots.$$

Finally, using our result (7.3) for the WKB spectrum, we get

$$\boxed{E_n^{(1)} = \frac{\pi^{2/3}}{3^{1/3}}\left(\frac{\Delta\alpha}{\alpha_0}\right)\left(\frac{\hbar^2\alpha_0^2}{m}\right)^{1/3}(n - 1/4)^{2/3}}$$

7.1.5 *A dimensional estimate for the perturbative correction to the spectrum*

Starting from the Bohr-Sommerfeld rule, use dimensional analysis to estimate the first order correction to the spectrum, $E_n^{(1)}$.

Solution: In the first order of the perturbation theory, we get

$$E_n^{(1)} = \Delta\alpha\Phi_n \,,$$

where Φ_n does not depend on $\Delta\alpha$. If the Schrödinger equation (7.4) can be solved using the WKB approximation, the spectrum would be a function of three parameters: $\tilde{\eta} \equiv (\hbar(n - 1/4))^2/2m$, α_0, and $\Delta\alpha$. However, if one wishes to determine Φ_n, the latter must be excluded. We get

– *The principal units—the units of length and the units of energy*:

$$[\mathcal{L}]\,, \quad [\mathcal{E}]\,;$$

– *The input parameters and their units*:

$$[\tilde{\eta}] = \left[\mathcal{L}^2\right][\mathcal{E}]$$
$$[\alpha_0] = [(1/\mathcal{L})][\mathcal{E}]\,;$$

– *The set of independent dimensionless parameters* $= \emptyset$;
– *The principal scales—the length scale and the energy scale, examples of*:

$$\mathcal{L} = \left(\frac{\hbar^2}{m\alpha_0}\right)^{1/3}$$

$$\mathcal{E} = \left(\frac{\hbar^2\alpha_0^2}{m}\right)^{1/3}\,;$$

– *Solution for the unknown*:

$$[\Phi_n] = [\mathcal{L}] \Rightarrow \Phi_n \sim \mathcal{L} = \left(\frac{\hbar^2}{m\alpha_0}\right)^{1/3}\,.$$

Thus,

$$\boxed{E_n^{(1)} \sim \left(\frac{\Delta\alpha}{\alpha_0}\right)\left(\frac{\hbar^2\alpha_0^2}{m}\right)^{1/3}(n - 1/4)^{2/3}}$$

7.1.6 A perturbative calculation of the shift of the energy levels under a small change in the coupling constant. The second order

(a) *Find the second order (in $\Delta\alpha$) perturbation theory correction, $E_n^{(2)}$, in the spectrum of the gravitational problem (7.4) due to the correction (7.5). Use the WKB expressions for the matrix elements of the relevant observables.*

Solution: The Fourier components of the coordinate read

$$x_{\tilde{m}}^{(CM)}(E) = \frac{1}{T(E)} \int_0^{T(E)} dt(v_0(E)t - \frac{1}{2}(\alpha_0/m)t^2) \exp[-i2\pi\tilde{m}t/T(E)]$$

$$= -\frac{2E}{\pi^2\tilde{m}^2},$$

where $v_0(E) = \sqrt{2E/m}$ is the initial velocity, and $T(E) = 2mv_0(E)/\alpha_0$. Now, the second order perturbation theory correction is

$$E_n^{(2)} = \sum_{n' \neq n} \left(E_n^{(2)}\right)_{n'},$$

where an individual term of the series reads[1]

$$\left(E_n^{(2)}\right)_{n'}$$

$$\stackrel{WKB}{\approx} \Delta\alpha^2 \frac{(\frac{1}{2}x_{n'-n}^{(CM)}(E_{n'}) + \frac{1}{2}x_{n'-n}^{(CM)}(E_n))^2}{E_n - E_{n'}}$$

$$= \left(\frac{\Delta\alpha}{\alpha_0}\right)^2 \left(\frac{\hbar^2\alpha_0^2}{m}\right)^{1/3} \frac{3^{2/3}\left((4n'-1)^{2/3} + (4n-1)^{2/3}\right)^2}{4\sqrt[3]{2}\pi^{10/3}(n'-n)^4\left((4n-1)^{2/3} - (4n'-1)^{2/3}\right)}$$

$$= \left(\frac{\Delta\alpha}{\alpha_0}\right)^2 \left(\frac{\hbar^2\alpha_0^2}{m}\right)^{1/3} \left\{ -\frac{3\left(3^{2/3}(4n-1)^{5/3}\right)}{8\left(\sqrt[3]{2}\pi^{10/3}\right)\Delta n^5} \right.$$

$$\left. -\frac{5(12n-3)^{2/3}}{4\left(\sqrt[3]{2}\pi^{10/3}\right)\Delta n^4} - \frac{7}{6\Delta n^3\left(\pi^{10/3}\sqrt[3]{24n-6}\right)} + \mathcal{O}\left(\frac{1}{\Delta n^2}\right)\right\}.$$

---

[1]Here, the convention for converting classical Fourier components to the quantum matrix elements is slightly different from the one suggested by the formulas (5.24)-(5.26). There, the classical Fourier componenets are taken at the mean energy $\bar{E}_{n',n} = (E_{n'} + E_n)/2$ between the two energies—$E_{n'}$ and E_n—involved in the $n' \leftrightarrow n$ transition; in the Problem 7.1.6 above, we are instead averaging the Fourier components themselves between their values at $E_{n'}$ and E_n. The difference between the two conventions is below the accuracy of the WKB estimates; however, the convention used here is slightly easier to handle.

Counterintuitively, it is the second $1/\Delta n^4$ term that gives the dominant contribution to the sum:

$$E_n^{(2)} = -\frac{\pi^{2/3}}{2 \times 3^{4/3}} \left(\frac{\Delta\alpha}{\alpha_0}\right)^2 \left(\frac{\hbar^2 \alpha_0^2}{m}\right)^{1/3} (n - 1/4)^{2/3}$$

The contribution of the first $1/\Delta n^5$ term,

$$\delta(E_n^{(2)}) = -\left(\frac{\Delta\alpha}{\alpha_0}\right)^2 \left(\frac{\hbar^2 \alpha_0^2}{m}\right)^{1/3} \frac{3 \cdot 3^{2/3} \left(\frac{1}{n}\right)^{7/3}}{4\pi^{10/3}} + \mathcal{O}\left(\left(\frac{1}{n}\right)^{10/3}\right),$$

is heavily impeded by the sign alternation across the sum.

(b) *Use the result of (7.3) to find $E_n^{(2)}$.*

<u>Solution:</u> Let us make a replacement $\alpha \to \alpha_0 + \Delta\alpha$ and expand E_n in powers of $\Delta\alpha$:

$$
\begin{aligned}
E_n &= \frac{(3\pi)^{2/3}}{2} \left(\frac{\hbar^2 (\alpha_0 + \Delta\alpha)^2}{m}\right)^{1/3} (n - 1/4)^{2/3} \\
&= \left(1 + \frac{\Delta\alpha}{\alpha_0}\right)^{2/3} \frac{(3\pi)^{2/3}}{2} \left(\frac{\hbar^2 \alpha_0^2}{m}\right)^{1/3} (n - 1/4)^{2/3} \\
&= \left(1 + \frac{2}{3}\left(\frac{\Delta\alpha}{\alpha_0}\right) - \frac{1}{9}\left(\frac{\Delta\alpha}{\alpha_0}\right)^2 + \cdots\right) \frac{(3\pi)^{2/3}}{2} \left(\frac{\hbar^2 \alpha_0^2}{m}\right)^{1/3} (n - 1/4)^{2/3} \\
&= E_n^{(0)} + \frac{2}{3}\left(\frac{\Delta\alpha}{\alpha_0}\right) E_n^{(0)} - \frac{1}{9}\left(\frac{\Delta\alpha}{\alpha_0}\right)^2 E_n^{(0)} + \cdots .
\end{aligned}
$$

On the other hand

$$E_n = E_n^{(0)} + E_n^{(1)} + E_n^{(2)} + \cdots .$$

Thus

$$
\begin{aligned}
E_n^{(2)} &= -\frac{1}{9}\left(\frac{\Delta\alpha}{\alpha_0}\right)^2 E_n^{(0)} \\
&= -\frac{\pi^{2/3}}{2 \times 3^{4/3}} \left(\frac{\Delta\alpha}{\alpha_0}\right)^2 \left(\frac{\hbar^2 \alpha_0^2}{m}\right)^{1/3} (n - 1/4)^{2/3}
\end{aligned}
$$

7.1.7 *A simple variational treatment of the ground state of a gravitational well*

Consider the motion of a mass m particle in a gravitational field plus a hard floor:

$$\hat{H} = -\frac{\hbar^2}{2m}\frac{\partial^2}{\partial x^2} + \alpha x \,,$$

or, in the dimensionless form

$$\hat{H} = -\frac{1}{2}\frac{\partial^2}{\partial x^2} + x \,.$$

Here, we have chosen a system of units where $\hbar = m = \alpha = 1$.

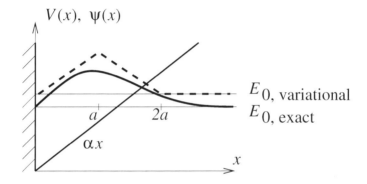

Fig. 7.2 A variational estimate for the ground state of the gravitational problem.

Find an approximate value for the ground state energy E_0 using the following variational ansatz:

$$\psi_0(x) = const \times \begin{cases} \dfrac{x}{a} \,, & 0 < x < a \\[2mm] 2 - \dfrac{x}{a} \,, & a < x < 2a \\[2mm] 0 \,, & 2a < x \end{cases} \qquad (7.6)$$

Compare with the exact value $E_0 = 1.8557$. Here, a is the only variational parameter (after normalization).

<u>Solution:</u>
First of all, let us normalize ψ_0 using $\int_0^{+\infty} dx\, |\psi_0(x)|^2 = 1$:

$$\psi_0(x) = \sqrt{\frac{3}{2a}} \times \begin{cases} \dfrac{x}{a} & , \quad 0 < x < a \\[2mm] 2 - \dfrac{x}{a} & , \quad a < x < 2a \\[2mm] 0 & , \quad 2a < x \end{cases} \cdot$$

The kinetic energy functional gives

$$T(a) = \int_0^{+\infty} dx\, \frac{1}{2} \left| \frac{\partial \psi_0}{\partial x} \right|^2 = \frac{3}{2a^2} \cdot$$

The potential energy gives

$$V(a) = \int_0^{+\infty} dx\, x\, |\psi_0|^2 = a \,.$$

The minimum of the total energy

$$E(a) = T(a) + V(a)$$

occurs at

$$a_{\min} = 3^{1/3} \,,$$

where the energy is given by

$$E_0 \approx E(a_{\min}) = \frac{3^{4/3}}{2} = 2.163\ldots$$

Note that this result corresponds to a relative error of 17%.

Chapter 8

Miscellaneous

8.1 Solved problems

8.1.1 A dimensional approach to the question of the number of bound states in δ-potential well ...

Consider the Schrödinger equation for a particle in a δ-well:

$$-\frac{\hbar^2}{2m}\frac{\partial^2}{\partial x^2}\psi(x) - g\delta(x)\psi(x) = E\psi(x)$$
$$g > 0\,.$$

Using dimensional analysis, show that the number of bound states is the same for any $g > 0$.

Solution: Assume that at some $g = g_N$, the number of bound states changes from N to $N+1$. Let's attempt to estimate g_N using dimensional analysis:

– *The principal units—the units of length and the units of energy*:

$$[\mathcal{L}]\,, \quad [\mathcal{E}]\,;$$

– *The* only *input parameter and its units*:

$$\left[\eta \equiv \hbar^2/m\right] = \left[\mathcal{L}^2\right][\mathcal{E}]\,;$$

– *The set of independent dimensionless parameters* $= \emptyset$;
– *The* only *principal scale, example of*:

$$\mathcal{H} = \frac{\hbar^2}{m}\,;$$

– No *solution for the unknown*:

$$[g_N] = [\mathcal{E}][\mathcal{L}] \Rightarrow \nexists\, g_N\,.$$

The problem is overdetermined, and thus such g_N does not exist. Hence, the number of bound states remains the same for all $g > 0$.

$$\boxed{Q.E.D.}$$

Remark: In reality, for $g > 0$, there is always one and only one bound state; its energy is $E_{g.s.} = -mg^2/(2\hbar^2)$.

8.1.2   *... and in a Pöschl-Teller potential*

Consider a Schrödinger equation for a particle in a *sech*^{2} potential well:

$$-\frac{\hbar^2}{2m}\frac{\partial^2}{\partial x^2}\psi(x) - \frac{\alpha}{\cosh^2(\kappa x)}\psi(x) = E\psi(x).$$

It is a general property of one-dimensional quantum mechanics, that any, no matter how shallow, potential well is able to support at least one bound state. In particular, if α in Eq. 8.1 is small, there is only one bound state. As one increases α to a certain threshold value α_1, another (second) bound state emerges.

Using dimensional analysis, estimate α_1.

Solution: The independent input parameters are $\hbar^2/m \equiv \eta$, α, and κ. Introduce a dimensionless parameter ν, such that $\lceil \nu \rceil$ equals the number of bound states. Here, $\lceil \xi \rceil$ is the ceiling function, the smallest integer not less than ξ. Units of the parameters of the problem are given by $[\eta] = [\mathcal{E}][\mathcal{L}]^2$, $[\alpha] = [\mathcal{E}]$, and $[\kappa] = [1/\mathcal{L}]$. It follows that

$$\nu \sim \Phi(\alpha/(\eta\kappa^2)),$$

where $\Phi(\xi)$ is a dimensionless function of a dimensionless argument that, for an argument ξ of the order of unity, assumes values of the order of unity.

The second bound state appears as soon as ν exceeds 1, where $\Phi(\alpha/(\eta\kappa^2)) \sim 1$. Thus, at this point, $\alpha/(\eta\kappa^2) \sim 1$. Hence

$$\boxed{\alpha_1 \sim \frac{\hbar^2\kappa^2}{m}}$$

Remark: The estimate above reproduces, entirely by accident, the exact result,

$$\alpha_1 = \frac{\hbar^2\kappa^2}{m}.$$

8.1.3 *Existence of lossless eigenstates in the* $1/x^2$-*potential*

Consider the Schrödinger equation for a particle in a $1/x^2$ potential:

$$-\frac{\hbar^2}{2m}\frac{\partial^2}{\partial x^2}\psi(x) - \frac{\alpha}{x^2}\psi(x) = E\psi(x)$$
$$x \geq 0\,.$$

It is known that there exists a critical value of the potential strength α, denoted as α^*, such that for $\alpha > \alpha^*$, there is no "physical" solution of the Schrödinger equation above, regardless of the energy E. Here, the "physicality" refers to the absence of any particle loss through the $x = 0$ boundary. More rigorously, for the physically relevant solutions, the probability current, $-j(x = 0)$ from the "physical region" $x \geq 0$ to the "unphysical region" $x < 0$ is zero: $-j(x = 0) = 0$, where

$$j(x) \equiv \frac{\hbar}{2mi}\left(\psi^*(x)\frac{\partial}{\partial x}\psi(x) - \psi(x)\frac{\partial}{\partial x}\psi^*(x)\right) \qquad (8.1)$$

is the probability current in the positive direction at point x.

Physically, the above transition corresponds to the opening of a loss channel, where particles start falling onto the center of the potential, with no return. Classically, this happens for any $\alpha > 0$. The quantum-mechanical falling-on-center threshold, α^*, is greater than zero however: in quantum mechanics, an attractive but sufficiently weak $1/x^2$-potential can still support "lossless" solutions.

(a) Consider an eigenstate $\psi(x)$, corresponding to an energy E. Assume that at short distances, the behavior of $\psi(x)$ is governed by a power law,

$$\psi(x) \overset{x\to 0+}{=} \text{const} \times x^\nu + o(x^\nu)\,,$$

where the exponent ν is unknown. As usual, $o(x^\nu)$ is a function $f(x)$ such that $\lim_{x\to 0+} f(x)/x^\nu = 0$. *Keeping only the dominant (at $x \to 0^+$) terms in Eq. 8.1, determine ν. Then analyze the probability current and find α^*;*

Solution: The dominant behavior at $x \to 0^+$ is governed by the $x^{\nu-2}$ terms. Assuming they cancel, we get

$$-\frac{\hbar^2}{2m}\nu(\nu-1)x^{\nu-2} - \alpha x^{\nu-2} + o(x^{\nu-2}) = o(x^{\nu-2})\,,$$

or

$$\nu_1 = \frac{1}{2}(1 + \sqrt{1 - 8\alpha/\eta})$$
$$\nu_2 = \frac{1}{2}(1 - \sqrt{1 - 8\alpha/\eta})\,,$$

where $\eta = \hbar^2/m$. For $\alpha > (1/8)\eta$, both solutions lead to a non-zero, finite flux, $j \propto$ const., that is able to transport particles between the physical $(x > 0)$ and unphysical $(x < 0)$ regions:

$$\psi_{1,2} \propto \sqrt{x}e^{\pm i\beta \ln(x)} \, ,$$

where $\beta = \sqrt{8\alpha/\eta - 1}$. For $\alpha < (1/8)\eta$ the second solution gives a diverging flux, $j \propto 1/x^{\sqrt{1-8\alpha/\eta}}$. The first however gives a vanishing flux: $j \propto x^{\sqrt{1-8\alpha/\eta}}$. Hence, $\alpha^\star = (1/8)\eta$, or

$$\boxed{\alpha^\star = \frac{\hbar^2}{8m}}$$

(b) *Estimate $\alpha^\star$ using dimensional analysis.*
<u>Solution:</u> The only dimensionful parameter in the problem is $\eta \equiv \hbar^2/m$. Its units are $[\eta] = [\mathcal{E}] [\mathcal{L}]^2$. $\alpha^\star$ is measured in the same units. Therefore:

$$\boxed{\alpha^\star \sim \frac{\hbar^2}{m}}$$

8.1.4 *On the absence of the unitary limit in two dimensions*

Consider the Schrödinger equations for the radially symmetric eigenstates in one, two, and three dimensions respectively,

$$-\frac{\hbar^2}{2m} \frac{\partial^2}{\partial r^2} \psi(r) = E\psi(r) \ \ 1D$$

$$-\frac{\hbar^2}{2m} \frac{1}{r} \frac{\partial}{\partial r} \left(r \frac{\partial}{\partial r} \right) \psi(r) = E\psi(r) \ \ 2D$$

$$-\frac{\hbar^2}{2m} \frac{1}{r^2} \frac{\partial}{\partial r} \left(r^2 \frac{\partial}{\partial r} \right) \psi(r) = E\psi(r) \ \ 3D \, ,$$

where $r = |x|$, $r = \sqrt{x^2 + y^2}$, and $r = \sqrt{x^2 + y^2 + z^2}$, in 1D, 2D, and 3D respectively. In 1D, even states, $\psi(-x) = \psi(x)$, play a role of the radially symmetric states. We will further assume a presence of a zero-range scatterer at the origin; the influence of the scatterer will be encoded in some nontrivial boundary conditions $r = 0^+$:

$$\psi(r) = \alpha + \beta r + \mathcal{O}(x^2) \quad 1D$$
$$\psi(r) = \alpha \ln(r) + \beta + \mathcal{O}(r \ln(r)) \quad 2D \qquad (8.2)$$
$$\psi(r) = \frac{\alpha}{r} + \beta + \mathcal{O}(r) \quad 3D.$$

A trivial boundary condition (the so-called "free-space" boundary condition) corresponding to no external potential at all, reads

$$\psi(r) = \alpha + \mathcal{O}(r^2) \quad 1D$$
$$\psi(r) = \beta + \mathcal{O}(r \ln(r)) \quad 2D \qquad (8.3)$$
$$\psi(r) = \beta + \mathcal{O}(r) \quad 3D.$$

Note that in 1D, the r-term corresponds to a $|x|$ singularity, and thus could not be present if no potential is there.

(a) In 1D and 3D, one can identify another type of boundary condition (the so-called "unitary limit" boundary condition) that is different from (8.3), but similar to (8.3), in that a parameter of units of length will not be introduced. *Find the "unitary limit" boundary condition in 1D and 3D and prove that any pair of α and β different from the "free-space" or "unitary limit" one will generate a length scale.*

Solution:

The "unitary limit" boundary conditions read

$$\boxed{\begin{aligned} \psi(r) &= \beta r + \mathcal{O}(r^2) \quad 1D \\ \psi(r) &= \frac{\alpha}{r} + \mathcal{O}(r) \quad 3D \end{aligned}}$$

If both α and β are finite, then a length scale (the so-called scattering length) can be constructed: $a_{1D} = -\alpha/\beta$ and $a_{3D} = -\alpha/\beta$ in both cases. Conventionally, the scattering length is defined as a position of the node of the wavefunction (8.2) if one assumes that the $\mathcal{O}(\cdot)$ terms can be neglected. Physically, this length corresponds to the node of an $E = 0$ eigenstate.

$$\boxed{Q.E.D.}$$

Remark: In 2D, $a_{2D} = \exp[-\beta/\alpha]$.

(b) *Using dimensional reasoning, show that the "unitary limit" boundary condition would not be possible in 2D.*

<u>Solution:</u> The natural candidate for a unitary limit would be

$$\psi(r) \overset{?}{=} \alpha \ln(r) + \mathcal{O}(r \ln(r)) \quad 2D.$$

However, in this case, we would not be able to cancel the units of length under the "$\ln(\ldots)$" sign.

$$\boxed{Q.E.D.}$$

Chapter 9

The Hellmann-Feynman Theorem

9.1 Solved problems

9.1.1 Lieb-Liniger model

Consider the Lieb-Liniger model[1] for a gas of one-dimensional δ-interacting bosons on a ring:

$$\hat{H} = \sum_{i=1}^{N} -\frac{\hbar^2}{2m}\frac{\partial^2}{\partial x_i^2} + \frac{1}{2}g\sum_{i=1}^{N}\sum_{\substack{j=1 \\ j \neq i}}^{N}\delta(x_i - x_j). \tag{9.1}$$

The wave function of the system, $\Psi(x_1, x_2, \ldots, x_N)$, obeys the periodic boundary condition:

$$\Psi(\ldots, x_i + L, \ldots) = \Psi(\ldots, x_i, \ldots).$$

Here, N is the number of particles, L is the circumference of the ring, m is the mass of the particles, and g is the coupling constant.

(a) *Using dimensional analysis, estimate the ground state energy, $E_{g.s.}$ in the first order of perturbation theory in powers of g; assume the thermodynamic limit: $N \to \infty$, $L \to \infty$, $N/L = n$, where n is the particle density;*

Solution: To first-order perturbation theory, the ground state energy per particle, $e \equiv E_{g.s.}/N$, must have the form $e = g\,\Xi$, where Ξ does not depend on g. The units of Ξ are $[\Xi] = 1/[\mathcal{L}]$. On the other hand, in the thermodynamic limit, $1/n$ remains the only parameter that is measured in

---

[1]E. H. Lieb and W. Liniger, Exact analysis of an interacting Bose gas. I. The general solution and the ground state, *Physical Review* **130**, 1605, (1963).

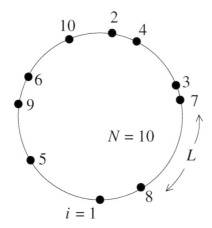

Fig. 9.1   Geometry of the Lieb-Liniger model with periodic boundary conditions.

units of length. Thus,

$$E_{\text{g.s.}} \sim gnN$$

The exact answer is

$$E_{\text{g.s.}} \approx \frac{1}{2} gnN \left(1 - \frac{4\sqrt{2}}{3\pi} \frac{1}{n|a_{1D}|} + \cdots \right),$$

where $a_{1D} = -2\hbar^2/mg$ is the so-called one-dimensional scattering length.

(b) *Using the Hellmann-Feynman theorem (see Sec. 9.3), prove that the ground state kinetic energy vanishes in the weak interaction limit. Use the result of sub-problem (a). Assume the thermodynamic limit.*

<u>Solution:</u> Introduce $\eta \equiv \hbar^2/m$. The derivative of the Hamiltonian (9.1) with respect to η gives

$$\frac{\partial}{\partial \eta} \hat{H} = \frac{1}{\eta} \hat{T},$$

where $\hat{T} \equiv \sum_i \hat{p}_i^2/(2m)$ is the kinetic energy. On the other hand,

$$\frac{\partial}{\partial \eta} E_{\text{g.s.}} = 0,$$

Thus

$$\langle g.s.|\hat{T}|g.s.\rangle \overset{g \to 0}{\to} 0$$

(c) *In the thermodynamic limit with strong interactions, the ground state energy becomes $E_{g.s.} = (\pi^2/6)(\hbar^2/m)\,nN$. Using the Hellmann-Feynman theorem, estimate the kinetic energy in this limit.*

Solution: Here,

$$\frac{\partial}{\partial \eta} E_{g.s.} = (\pi^2/6)\,nN\,,$$

Thus

$$\langle g.s.|\hat{T}|g.s.\rangle = E_{g.s.} = (\pi^2/6)(\hbar^2/m)\,nN$$

Remark: Alternatively, this result could have been obtained using the Hellmann-Feynman theorem with the coupling constant g—the prefactor in front of the interaction energy in the Hamiltonian (9.1)—as a parameter. Since the $g \to \infty$ ground state energy $E_{g.s.} = (\pi^2/6)(\hbar^2/m)\,nN$ does not contain the coupling constant g at all, the interaction energy should vanish in this limit, and the kinetic energy becomes equal to the total.

9.1.2 Expectation values of $1/r^2$ and $1/r$ in the Coulomb problem, using the Hellmann-Feynman theorem

The Hamiltonian for the Coulomb problem is

$$\hat{H} = -\frac{\hbar^2}{2m}\Delta_3 - \frac{\alpha}{r}\,. \tag{9.2}$$

The corresponding Hamiltonian for the radial motion reads

$$\hat{H}_r = \frac{\hat{p}_r^2}{2m} - \frac{\alpha}{r} + \frac{\hbar^2 l(l+1)}{2mr^2}\,, \tag{9.3}$$

where $\hat{p}_r \equiv -i\hbar\partial/\partial r$ (in the radial motion representation, $\chi = \Psi/r$). The spectrum is given by

$$E_{n,l,m} = -\frac{\alpha/a}{2n^2} = -\frac{\alpha/a}{2(n_r + l + 1)^2}$$
$$n = 1,\, 2,\, 3,\, \ldots$$
$$l = 0,\, 1,\, 2,\, \ldots\, n - 1$$
$$n_r = n - l - 1\,,$$

where $a = \hbar^2/(m\alpha)$ is the Bohr radius.

Assignment: by applying the Hellmann-Feynman theorem to the Hamiltonian (9.3) and its spectrum

$$E_{n_r} = -\frac{\alpha/a(\alpha)}{2(n_r + l + 1)^2} \tag{9.4}$$

$$n_r = 0, 1, 2, \dots .$$

(and regarding l and α as parameters) determine the expectation value of $1/r^2$ and $1/r$ for all values of n_r and l.

Solution:

Hellmann-Feynman theorem:

$$\langle n_r, l | \frac{\partial \hat{H}_r}{\partial l} | n_r, l \rangle = \frac{\partial E_{n_r}}{\partial l}.$$

The left hand side gives

$$\langle n_r, l | \frac{\partial \hat{H}_r}{\partial l} | n_r, l \rangle = \langle n_r, l | \frac{\hbar^2(2l+1)}{2mr^2} | n_r, l \rangle$$

$$= \frac{\hbar^2(2l+1)}{2m} \langle n_r, l | \frac{1}{r^2} | n_r, l \rangle.$$

In turn, the right hand side is

$$\frac{\partial E_{n_r}}{\partial l} = \frac{\alpha/a}{(n_r + l + 1)^3}.$$

Combining the two we get

$$\boxed{\langle n_r, l | \frac{1}{r^2} | n_r, l \rangle = \frac{1}{(n_r + l + 1)^3(l + \frac{1}{2})} \frac{1}{a^2}}$$

In the case of $n_r = 0$ and $l = 0$, the above relationship gives

$$\boxed{\langle n = 1, l = 0, m = 0 | \frac{1}{r^2} | n = 1, l = 0, m = 0 \rangle = \frac{2}{a^2}}$$

Likewise, for $1/r$ we get

$$\langle n_r, l | \frac{\partial \hat{H}_r}{\partial \alpha} | n_r, l \rangle = \frac{\partial E_{n_r}}{\partial \alpha}.$$

The left hand side:

$$\langle n_r, l | \frac{\partial \hat{H}_r}{\partial \alpha} | n_r, l \rangle = -\langle n_r, l | \frac{1}{r} | n_r, l \rangle.$$

The right hand side:

$$\frac{\partial E_{n_r}}{\partial \alpha} = -\frac{1/a}{(n_r + l + 1)^2} \,.$$

Combining the two we get

$$\langle n_r, l | \frac{1}{r} | n_r, l \rangle = \frac{1}{(n_r + l + 1)^2} \frac{1}{a}$$

9.1.3 Expectation value of the trapping energy in the ground state of the Calogero system

Consider the Hamiltonian for the Calogero system—N harmonically trapped bosonic particles interacting via a $1/x^2$ potential:[2]

$$\hat{H} = \hat{T} + \hat{V}_{\text{trap}} + \hat{U}_{\text{interaction}} \,,$$

where

$$\hat{T} = -\frac{\hbar^2}{2m} \sum_{i=1}^{N} \frac{\partial^2}{\partial x_i^2}$$

$$\hat{V}_{\text{trap}} = \frac{m\omega^2}{2} \sum_{i=1}^{N} x_i^2$$

$$\hat{U}_{\text{interaction}} = \frac{\hbar^2 \nu(\nu - 1)}{2m} \sum_{i=1}^{N} \sum_{\substack{j=1 \\ j \neq i}}^{N} \frac{1}{(x_i - x_j)^2} \,.$$

The ground state energy reads

$$E_0 = \hbar\omega \left(\frac{N}{2} + \frac{\nu N(N-1)}{2} \right).$$

Using the Hellmann-Feynman theorem, find the expectation value of the trapping energy, $\hat{V}_{trap}$, in the ground state.

Solution:

The Hellmann-Feynman theorem gives:

$$\langle \Psi_0 | \frac{\partial \hat{H}}{\partial \omega} | \Psi_0 \rangle = \frac{\partial}{\partial \omega} E_0 \,,$$

---

[2]F. Calogero, Ground state of one-dimensional N-body system, J. Math. Phys., **10**, 2197 (1969).

where $|\Psi_0\rangle$ is the ground state wavefunction. The left hand side reads

$$\langle\Psi_0|\frac{\partial\hat{H}}{\partial\omega}|\Psi_0\rangle = \frac{2}{\omega}\langle\Psi_0|\hat{V}_{\text{trap}}|\Psi_0\rangle\,.$$

In turn, the right hand side reads

$$\frac{\partial E_0}{\partial\omega} = \frac{1}{\omega}E_0\,.$$

Combining the two we get

$$\boxed{\langle\Psi_0|\hat{V}_{\text{trap}}|\Psi_0\rangle = \frac{1}{2}E_0}$$

9.1.4 *Virial theorem from the Hellmann-Feynman theorem*

Let us formulate a quantum version of the

<u>Virial theorem:</u>
Let

$$\hat{H} = -\frac{\hbar^2\Delta_{\vec{r}}}{2m} + V(\vec{r})$$

be a (generally multi-dimensional) Hamiltonian and assume that its potential is scale-invariant: there exists a parameter κ such that for any scale parameter λ the following relationship holds:

$$V(\lambda\vec{r}) = \lambda^\kappa V(\vec{r})\,. \tag{9.5}$$

Here

$$\Delta_{\vec{r}} \equiv \sum_{i=1}^{d}\frac{\partial^2}{\partial x_i^2}$$

is the Laplacian with respect to d coordinates $\vec{r} = (x, y, \ldots) = (x_1, x_2, \ldots)$, and d is the number of spatial dimensions.
Let $|\psi(\vec{r})\rangle$ be a bound state of this Hamiltonian, of an energy E:

$$\hat{H}|\psi\rangle = E|\psi\rangle \tag{9.6}$$

$$\langle\psi|\psi\rangle \equiv \int d^d\vec{r}\,|\psi(\vec{r})|^2 = 1\,. \tag{9.7}$$

Then the expectation values of the kinetic and potential energies in the state $\psi(\vec{r})$,

$$\hat{T} \equiv -\frac{\hbar^2\Delta}{2m}$$
$$\hat{V} \equiv V(\vec{r})\,,$$

are related by

$$2\langle\psi|\hat{T}|\psi\rangle = \kappa\langle\psi|\hat{V}|\psi\rangle \, . \tag{9.8}$$

In particular, for one-dimensional power-law potentials,

$$\hat{H} = -\frac{\hbar^2}{2m}\frac{\partial^2}{\partial x^2} + K_q x^{2q}$$
$$q = 1, 2, 3, \ldots \, ,$$

we have

$$\kappa = 2q \, ,$$

and thus

$$\langle\psi|\hat{T}|\psi\rangle = q\langle\psi|\hat{V}|\psi\rangle \, .$$

Show that the virial theorem follows from the Hellmann-Feynman theorem.

<u>Solution:</u> Let us rewrite the Schrödinger equation (9.6) in a coordinate form:

$$-\frac{\hbar^2\Delta}{2m}\psi(\vec{r}) + V(\vec{r})\psi(\vec{r}) = E\psi(\vec{r}) \, . \tag{9.9}$$

Consider a replacement of variables:

$$\vec{r} = \lambda\vec{r}' \, .$$

The Schrödinger equation (9.9) becomes

$$-\frac{1}{\lambda^2}\frac{\hbar^2\Delta_{\vec{r}'}}{2m}\psi(\lambda\vec{r}') + V(\lambda\vec{r}')\psi(\lambda\vec{r}') = E\psi(\lambda\vec{r}')$$

or

$$-\frac{1}{\lambda^2}\frac{\hbar^2\Delta_{\vec{r}'}}{2m}\psi(\lambda\vec{r}') + \lambda^\kappa V(\vec{r}')\psi(\lambda\vec{r}') = E\psi(\lambda\vec{r}') \, , \tag{9.10}$$

where we used the scaling property (9.5). Here, $\Delta_{\vec{r}'}$ is the Laplacian with respect to the coordinates $\vec{r}'$. Introduce the function

$$\psi_\lambda(\vec{r}') \equiv \sqrt{\lambda^d}\,\psi(\lambda\vec{r}') \, .$$

Observe that according to Eq. (9.10), the state $|\psi_\lambda\rangle$ is an eigenstate of the Hamiltonian

$$\hat{H}_\lambda = -\frac{1}{\lambda^2}\frac{\hbar^2\Delta_{\vec{r}'}}{2m} + \lambda^\kappa V(\vec{r}') \, , \tag{9.11}$$

corresponding to the eigenvalue E. This state is normalized to unity. Observe also that the eigenenergy E does *not* depend on the parameter λ.

At $\lambda = 1$, the Hamiltonian $\hat{H}_\lambda$ and its eigenstate $\psi_\lambda(\vec{r}')$ reduce to the Hamiltonian-eigenstate pair of the original problem.

According to the Hellmann-Feynman theorem,

$$\langle\psi_\lambda|\frac{d\hat{H}(\lambda)}{d\lambda}|\psi_\lambda\rangle = \frac{dE_{\vec{n}}(\lambda)}{d\lambda},$$

and in particular

$$\langle\psi_\lambda|\frac{d\hat{H}_\lambda}{d\lambda}|\psi_\lambda\rangle\Big|_{\lambda=0} = \frac{dE_{\vec{n}}(\lambda)}{d\lambda}\Big|_{\lambda=0}.$$

This leads to

$$\langle\psi_\lambda|(-2)\left\{-\frac{\hbar^2\Delta}{2m}\right\} + \kappa V(\vec{r})|\psi_\lambda\rangle = 0.$$

$$\boxed{Q.E.D.}$$

9.2 Problems without provided solutions

9.2.1 *Virial theorem for the logarithmic potential and its corollaries*

Consider a logarithmic potential again:

$$-\frac{\hbar^2}{2m}\frac{\partial^2}{\partial x^2}\psi(x) + 2\epsilon\ln(x/a)\psi(x) = E\psi(x). \tag{9.12}$$

(a) *Using the Hellmann-Feynman theorem, derive a quantum virial theorem for this potential.*

Answer: $\langle\psi|\hat{T}|\psi\rangle = \epsilon^3$.

(b) *Starting from your result in (a) and using the Hellmann-Feynman theorem once again, prove that the distances between the energy levels of (9.12) do not depend on the particle's mass:*[4]

$$\frac{\partial}{\partial m}(E_{n'} - E_n) = 0$$
for any n, n'.

---

[3] R. Mack, J. P. Dahl, H. Moya-Cessa, W. T. Strunz, R. Walser, and W. P. Schleich, Riemann ζ function from wave-packet dynamics, *Physical Review A* **82**, 032119 (2010).

[4] See "Surprises with Logarithm Potential" by Debnarayan Jana, `https://web.archiv e.org/web/20161019125550/http://physics.unipune.ernet.in/~phyed/27.3/1379(27 .3).pdf`. The derivation there is very brief; a more detailed proof is required.

9.3 Background

9.3.1 *The Hellmann-Feynman theorem*

Hellmann-Feynman theorem (Hellmann (1937); Feynman (1939)) states: let a Hamiltonian $\hat{H}(\lambda)$ (and, consequently, its eigenvalues $E_{\vec{n}}(\lambda)$ and eigenstates $|\psi_{\vec{n}}(\lambda)\rangle$) depend on a parameter λ. Then

$$\langle\psi_{\vec{n}}(\lambda)|\frac{d\hat{H}(\lambda)}{d\lambda}|\psi_{\vec{n}}(\lambda)\rangle = \frac{dE_{\vec{n}}(\lambda)}{d\lambda}. \tag{9.13}$$

Here, $\vec{n}$ is a set of quantum numbers that determine the eigenstate $|\psi_{\vec{n}}(\lambda)\rangle$.

Chapter 10

Local Density Approximation Theories

10.1 Solved problems

10.1.1 A Thomas-Fermi estimate for the atom size and total ionization energy

Minimization of the Thomas-Fermi energy functional for a neutral atom,

$$\mathcal{E}[n(\cdot)] \equiv \int d^3\vec{r} \left\{ \frac{\int_0^{p \leq p_F(n(\vec{r}))} d^3\vec{p} \left\{ \frac{p^2}{2m} \right\}}{\int_0^{p \leq p_F(n(\vec{r}))} d^3\vec{p} \ 1} - \frac{Ze^2}{r} \right\} n(\vec{r})$$

$$+ \frac{1}{2} \int d^3\vec{r} d^3\vec{r}' \frac{e^2}{|\vec{r} - \vec{r}'|} n(\vec{r}) n(\vec{r}') , \tag{10.1}$$

subject to an electron number constraint

$$\mathcal{N}[n(\cdot)] \equiv \int d^3\vec{r} \, n(\vec{r}) = Z , \tag{10.2}$$

with the density distribution $n(\vec{r})$ as the variational field, leads to the following equation for the electron density:

$$\frac{p_F^2(\vec{r})}{2m} - \frac{Ze^2}{r} + \int d^3\vec{r}' \frac{e^2}{|\vec{r} - \vec{r}'|} n(\vec{r}') = 0 , \tag{10.3}$$

where $p_F(n) \equiv (3\pi^2)^{1/3} \hbar n^{1/3}$ is the local Fermi momentum, $n(\vec{r})$ is the electron density, $e < 0$ is the electron charge, m is the electron mass, and Z is the atomic number (number of protons and—equal to it—number of electrons). The Fermi momentum chosen this way guarantees that the phase space density $\sigma(\vec{r}, \vec{p})$ in all points of phase space occupied by electrons will be equal to $\sigma(\vec{r}, \vec{p}) = 2/(2\pi\hbar)^3$.

Using dimensional analysis, estimate the typical distance from the nucleus that electrons reside at (say the radius of a sphere containing half of *all electrons) and the total ionization energy (the minimal energy needed to release all the electrons from the atom). As usual, try to incorporate the dimensionless parameter Z into some dimensionful parameters.*

No prior knowledge of the Thomas-Fermi theory is assumed.

<u>Solution:</u> Introduce the probability density for an individual electron,

$$w(\vec{r}) \equiv \frac{n(\vec{r})}{Z}.$$

Then, Eq. (10.3) becomes

$$\frac{(3\pi^2)^{2/3}}{2} R w(\vec{r})^{2/3} - \frac{1}{r} + \int d^3\vec{r}\, \frac{1}{|\vec{r} - \vec{r}'|} w(\vec{r}') = 0,$$

where the radius R is given by

$$R = \frac{1}{Z^{1/3}} \frac{\hbar^2}{me^2}$$

Since it is the only length scale in the problem, it indeed determines the typical distance between an electron and the nucleus.

The total ionization energy we estimate as the electron-nucleus energy $E_{\text{ionization}} \sim Z \times Ze^2/R$:

$$E_{\text{ionization}} \sim Z^{7/3} \frac{me^4}{\hbar^2}$$

10.1.2   *The size of an ion*

Neglecting the interaction between electrons, estimate the size of a positive ion that is obtained by stripping M elecrons from a neutral atom of atomic number Z.

Hint: Recall which length was giving the length scale for the size of a neutral atom: use the same logic here.

<u>Solution:</u> The size of an ion will be given by the radius of the outer orbit:

$$R_{\text{ion}} \sim \langle r \rangle_{\text{outer}}.$$

Fig. 10.1   Total ionization energy: real atoms vs. Thomas-Fermi theory.

On the other hand, the radius of any orbit is given by

$$\langle r \rangle_{n,\,l,\,m} \sim \frac{n^2}{Z}\,a_{\mathrm{B}}\,,$$

and in particular

$$\langle r \rangle_{\mathrm{outer}} \sim \frac{(n_{\mathrm{outer}})^2}{Z}\,a_{\mathrm{B}}\,.$$

Also, the principal quantum number n scales as the cube root of the number of states below n. For the ground state of ion the latter is just the number of electrons, $Z - M$:

$$n_{\mathrm{outer}} \sim (Z - M)^{\frac{1}{3}}\,.$$

Combining all of the above, we get

$$\boxed{R_{\mathrm{ion}} \sim \frac{(Z - M)^{\frac{2}{3}}}{Z}\,a_{\mathrm{B}}}$$

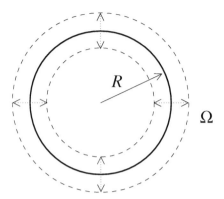

Fig. 10.2   A cloud of cold bosons and its breathing excitation.

10.1.3   *Time-dependent Thomas-Fermi model for cold bosons*

Consider the time-dependent Thomas-Fermi model for three-dimensional harmonically trapped short-range-interacting bosons:

$$\begin{cases} \dfrac{\partial}{\partial t}n + \vec{\nabla}(n\vec{v}) = 0 \\[2ex] \dfrac{\partial}{\partial t}\vec{v} + (\vec{v} \cdot \vec{\nabla})\vec{v} = -\dfrac{1}{m}g\vec{\nabla}n - \omega^2\vec{r} \\[2ex] \int d^3\vec{r}\,n = N \end{cases},$$

where $\vec{r} = (x,\, y,\, z)$ is a three-dimensional coordinate, $n = n(\vec{r})$ is atomic density, $\vec{v} = \vec{v}(\vec{r})$ is the local mean velocity of atoms, ω is the trapping frequency, g is the interatomic interaction strength, and m is the atomic mass. The first equation above ensures the *continuity* of the atomic flow. The second equation, the so-called *Euler* equation, constitutes Newton's second law for an individual atom, subject to a pressure force from the other atoms, $\vec{f}_{\text{press.}} = -g\vec{\nabla}n$, and a trapping force, $\vec{f}_{\text{trap}} = -\omega^2\vec{r}$. The third relationship fixes the number of atoms to N.

Using dimensional analysis, estimate the size of the atomic cloud in the ground state of the system, R, and the frequency of the lowest breathing excitation, Ω (see Figure 10.2).

<u>Solution:</u> Introducing the one-body probability density, $w \equiv n/N$, we get

$$
\begin{cases}
\dfrac{\partial}{\partial t} w + \vec{\nabla}(w\vec{v}) = 0 \\[2mm]
\dfrac{\partial}{\partial t} \vec{v} + (\vec{v} \cdot \vec{\nabla})\vec{v} = -\Lambda\vec{\nabla}w - \omega^2 \vec{r} \\[2mm]
\int d^3\vec{r}\, w = 1
\end{cases}
$$

where $\Lambda \equiv gN/m$. Then

– *The principal units—the units of length and the units of time:*
$$[\mathcal{L}], \quad [\mathcal{T}];$$
– *The input parameters and their units:*
$$[\omega] = 1/[\mathcal{T}]$$
$$[\Lambda] = [\mathcal{L}]^5 / [\mathcal{T}]^2 ;$$
– *The set of independent dimensionless parameters* $= \emptyset$;
– *The principal scales—the length scale and the time scale, examples of:*
$$\mathcal{L} = \left(\frac{\Lambda}{\omega^2}\right)^{1/5} = \left(\frac{gN}{m\omega^2}\right)^{1/5}$$
$$\mathcal{T} = \frac{1}{\omega};$$
– *Solution for the unknowns:*
$$[R] = [\mathcal{L}] \Rightarrow R \sim \mathcal{L}$$
$$[\Omega] = 1/[\mathcal{T}] \Rightarrow \Omega \sim 1/\mathcal{T}.$$

Therefore,

$$\boxed{\begin{aligned} R &\sim \left(\frac{gN}{m\omega^2}\right)^{1/5} \\ \Omega &\sim \omega \end{aligned}}$$

Remark: The exact result is
$$R = \left(\frac{15}{4\pi}\frac{gN}{m\omega^2}\right)^{1/5}$$
$$\Omega = \sqrt{5}\,\omega$$

10.2 Problems without provided solutions

10.2.1 *The quantum dot*

Consider Z non-interacting electrons of mass m each in a quantum dot. The corresponding energy-vs-density functional reads

$$\mathcal{E}[n(\cdot)] = T[n(\cdot)] + V[n(\cdot)] \,,$$

where

$$T[n(\cdot)] = \frac{3}{10}(3\pi^2)^{2/3}\frac{\hbar^2}{m}\int d^3\vec{r}\,(n(\vec{r}))^{5/3}$$

is the kinetic energy (the same as in the atomic problem), and

$$V[n(\cdot)] = \frac{m\omega^2}{2}\int d^3\vec{r}\,r^2 n(\vec{r})$$

is the confinement energy. Using dimensional reasoning, estimate the size of the electron cloud in the ground state of the dot.

10.2.2 *Dimensional analysis of an atom beyond the Thomas-Fermi model*

Consider the *exact* Schrödinger equation for electrons in an atom:

$$\hat{H}\Psi = E\Psi\,,$$

where

$$\Psi = \Psi(\vec{r}_1, \vec{r}_2, \ldots, \vec{r}_Z)\,,$$

is the wavefunction for Z three-dimensional electrons, and

$$\hat{H} \equiv \sum_{i=1}^{Z} -\frac{\hbar^2}{2m}\Delta_{\vec{r}_i} - \sum_{i=1}^{Z}\frac{Ze^2}{r_i} + \sum_{i=1}^{Z-1}\sum_{j=i+1}^{Z}\frac{e^2}{|\vec{r}_i - \vec{r}_j|}\,,$$

where m is the electron mass and $-|e|$ is the electron charge.

Show that at the level of the exact equations, an up-to-a-factor dimensional estimate of the ionization energy is not possible. Use the counting scheme (1.5).

Chapter 11

Integrable Partial Differential Equations

11.1 Solved problems

11.1.1 *Solitons of the Korteweg-de Vries equation*

Consider the Korteweg-de Vries (KdV) equation,

$$\frac{\partial}{\partial t}u + 6u\frac{\partial}{\partial x}u + \frac{\partial^3}{\partial x^3}u = 0, \qquad (11.1)$$

where $u = u(x,t)$ is a function of coordinate and time. The KdV equation is known to possess solitonic solutions,[1]

$$u(x,\,t) = u_0\,\text{sech}^2[(x - vt)/\Delta x],$$

where v is the soliton velocity, x_0 is its initial position, u_0 is its amplitude, and Δx is its width.

Using dimensional analysis only, estimate the width and the amplitude of the soliton.

Solution: Assume that x has units of length, and t has units of time. Regardless of what we chose for the units of u, in Eq. 11.1 the first and last terms on the left hand side are not mutually consistent. However, let us temporarily introduce a constant η in front of the third derivative:

$$\frac{\partial}{\partial t}u + 6u\frac{\partial}{\partial x}u + \eta\frac{\partial^3}{\partial x^3}u = 0. \qquad (11.2)$$

The units of measurement for η must be given by $[\mathcal{L}]^3/[\mathcal{T}]$, where $[\mathcal{L}]$ and $[\mathcal{T}]$ are units of length and time respectively. According to the same

---

[1]In fact, these are the first solitons ever discovered, see J. Scott Russell 1844, Report on waves, *Report of the Fourteenth Meeting of the British Association for the Advancement of Science, Held at York in September 1844* (John Murray, Albemarle Street, London, 1845), p. 311; D. J. Korteweg & G. de Vries, On the change of form of long waves advancing in a rectangular canal, and on a new type of long stationary waves, *Philosophical Magazine*, 5th ser., **39**, 422 (1895).

equation, the units for the unknown field $u(x, t)$ are the ones for velocity: $[u] = [\mathcal{L}] / [\mathcal{T}]$. Obviously, the soliton velocity v is measured in the same units.

Now, the problem is reduced to the traditional form of physics problems. The analysis that follows is straightforward.

– *The principal units—the units of length and the units of time:*

$$[\mathcal{L}], \quad [\mathcal{T}];$$

– *The input parameters and their units:*

$$[\eta] = [\mathcal{L}]^3 / [\mathcal{T}]$$
$$[v] = [\mathcal{L}] / [\mathcal{T}];$$

– *The set of independent dimensionless parameters* $= \emptyset$;
– *The principal scales—the length scale and the time scale, examples of:*

$$\mathcal{L} = \left(\frac{\eta}{v}\right)^{1/2}$$
$$\mathcal{T} = \left(\frac{\eta}{v^3}\right)^{1/2};$$

– *Solution for the unknowns:*

$$[\Delta x] = [\mathcal{L}] \Rightarrow \Delta x \sim \mathcal{L} = \left(\frac{\eta}{v}\right)^{1/2}$$
$$[u_0] = [\mathcal{L}] / [\mathcal{T}] \Rightarrow u_0 \sim \mathcal{L}/\mathcal{T} = v.$$

Now we can retreat to the dimensionless form (11.1) of the KdV equation and set η to unity. We get

$$\boxed{\begin{aligned} \Delta x &\sim \frac{1}{\sqrt{v}} \\[2ex] u_0 &\sim v \end{aligned}}$$

Remark: The exact values are

$$\Delta x = \frac{2}{\sqrt{v}}$$
$$u_0 = \frac{v}{2}.$$

11.1.2 *Breathers of the nonlinear Schrödinger equation*

Consider a dimensionless form of the nonlinear Schrödinger equation (NLSE):

$$i\frac{\partial}{\partial t}u = -\frac{\partial^2}{\partial x^2}u - 2|u|^2 u. \tag{11.3}$$

It is known to possess a family of so-called "breather"[2] solutions (localized time-periodic excitations):

$$u_L(x, t) = u_0(L)f(x/L, t/T(L))$$
$$u_L(x, t + T(L)) = u_L(x, t).$$

Members of the family are parametrized by their size L.

 Show that the "breathing" period is proportional to the square of the size:

$$T(L) \propto L^2.$$

Solution: Introduce temporarily a constant η:

$$i\frac{\partial}{\partial t}u = -\eta\frac{\partial^2}{\partial x^2}u - 2|u|^2 u.$$

Let us measure x and t in units of length and time respectively. Then, the units for η become $[\eta] = [\mathcal{L}^2]/[\mathcal{T}]$. The only other dimensionful parameter of the problem is the size of the breather L. Thus, the period T becomes

$$T \sim \frac{L^2}{\eta}.$$

Now we can return η to its unit value, $\eta \to 1$: we finally get

$$\boxed{T \propto L^2}$$

Remark: The actual breathers have a form

$$u(x,t) = A' \times \frac{4e^{it'}\left(\cosh(3x') + 3e^{8it'}\cosh(x')\right)}{3\cos(8t') + 4\cosh(2x') + \cosh(4x')},$$

where[3] $A' = \xi$, $x' = \xi x$, and $t = \xi^2 t$.

---

[2] Rigoriusly speaking, the nonlinear Schrödinger breathers—unlike the paradigmatic breathers of the sine-Gordon equation—are not stable, and given a small perturbation, decay into several solitons. Nevertheless, if perturbation of the initial state is sufficiently small or absent, then the breathing behavior will be visible for at least a few periods.

[3] D. Schrader, Explicit calculation of N-soliton solutions of the nonlinear Schroedinger equation, *IEEE Journal of Quantum Electronics* **31**, 2221 (1995).

11.1.3 *Healing length*

Consider a one-dimensional nonlinear Schrödinger equation

$$-\frac{\hbar^2}{2m}\frac{\partial^2}{\partial x^2}\psi(x) + g|\psi(x)^2|\psi(x) = \mu\psi(x),$$

where m is the atomic mass and $g > 0$ is the coupling constant. This equation describes the motion of interacting quantum Bose gases, the dynamics of the superfluid Helium-4, and the propagation of light through nonlinear fibers. We will also assume that the $x < 0$ half-space is filled by an impenetrable wall:

$$\psi(x) = 0, \quad \text{for } x < 0.$$

Unlike the other problems in this Chapter, we will retain physical units in the equation we are analyzing for this problem; the healing length is a concept of broad physical significance, whose usefulness extends far beyond the integrable one-dimensional case: in particular the healing length provides an estimate for the size of the core of a superfluid vortex.

Consider a solution that produces a constant density n_0 far away from the wall:

$$\psi(x) \approx \sqrt{n_0} = \text{const.}, \quad \text{for } x \to +\infty. \tag{11.4}$$

Using dimensional analysis, estimate the healing length ξ, such that for $x \gtrsim \xi$, the density $n(x) \equiv |\psi(x)|^2$ already approaches its limiting value n_0.

Solution: Using the boundary condition (11.4), the chemical potential assumes a value of $\mu = gn_0$. Thus, the independent input parameters are $\hbar^2/m \equiv \eta$, g, and n_0. The dimensional estimate for ξ is as follows.

– *The principal units—the units of length and the units of energy*:

$$[\mathcal{L}], \quad [\mathcal{E}];$$

– *The input parameters and their units*:

$$[\eta] = [\mathcal{E}][\mathcal{L}]^2$$
$$[g] = [\mathcal{E}][\mathcal{L}]$$
$$[n_0] = 1/[\mathcal{L}];$$

– *The set of independent dimensionless parameters =*

$$\left\{P_1 \equiv \eta n_0/g\right\};$$

– *The principal scales—the length scale and the energy scale, examples of*:

$$\mathcal{L} = 1/n_0$$
$$\mathcal{E} = g^2/\eta \; ;$$

– *Solution for the unknown*:

$$[\xi] = [\mathcal{L}] \Rightarrow \xi = \Phi(P_1) \times \mathcal{L} = \Phi(\eta n_0/g) \times (1/n_0) \,,$$

where $\Phi(P)$ is an arbitrary function of one variable.

So far, it looks like an underdetermined problem. However, if we introduce a field $\phi(x)$, such that

$$\phi(x) \equiv \psi(x)/\sqrt{n_0} \,.$$

It obeys

$$-\frac{\hbar^2}{2m}\frac{\partial^2}{\partial x^2}\phi(x) + gn_0|\phi(x)^2|\phi(x) = gn_0\phi(x)$$
$$\phi(x) = 0, \quad \text{for } x < 0$$
$$\phi(x) \approx 1, \quad \text{for } x \to +\infty \,.$$

Note that $\phi(x)$ will have the same healing length ξ.

Now, the set of independent parameters has only two members: $\hbar^2/m \equiv \eta$ and $gn_0 \equiv G$. The healing length ξ can thus be found as follows:

– *The principal units—the units of length and the units of energy*:

$$[\mathcal{L}], \quad [\mathcal{E}] \; ;$$

– *The input parameters and their units*:

$$[\eta] = [\mathcal{E}][\mathcal{L}]^2$$
$$[G] = [\mathcal{E}] \; ;$$

– *The set of independent dimensionless parameters* $= \emptyset$;
– *The principal scales—the length scale and the energy scale, examples of*:

$$\mathcal{L} = \sqrt{\eta/G} = \hbar/\sqrt{mgn_0}$$
$$\mathcal{E} = G = gn_0 \; ;$$

– *Solution for the unknown*:

$$[\xi] = [\mathcal{L}] \Rightarrow \xi \sim \mathcal{L} = \hbar/\sqrt{mgn_0} \,.$$

Thus

$$\xi \sim \frac{\hbar}{\sqrt{mgn_0}}$$

Remark: the exact solution reads

$$\psi(x) = \begin{cases} 0, & \text{for } x < 0 \\ \sqrt{n_0}\tanh(x/\xi), & \text{for } x \geq 0 \end{cases},$$

with

$$\xi = \frac{\hbar}{\sqrt{mgn_0}}.$$

11.1.4 *Dimensional analysis of the projectile problem as a prelude to a discussion on the Kadomtsev-Petviashvili solitons*

Consider Newton's equations for a projectile,

$$\ddot{x} = 0$$

$$\ddot{y} = -g \qquad\qquad (11.5)$$

$$x\big|_{t=0} = 0 \qquad \dot{x}\big|_{t=0} = v_{x,0}$$

$$y\big|_{t=0} = 0 \qquad \dot{y}\big|_{t=0} = v_{y,0}.$$

(a) *Using straightforward dimensional analysis, try to determine the horizontal distance L_x traveled by the projectile before it falls to the ground $(y = 0)$. Naively, the problem would seem underdetermined: no answer will ensue.*

Solution:

– *The principal units—the units of length and the units of time*:

$$[\mathcal{L}], \quad [\mathcal{T}];$$

– *The input parameters and their units*:

$$[v_{x,0}] = [\mathcal{L}]/[\mathcal{T}]$$
$$[v_{y,0}] = [\mathcal{L}]/[\mathcal{T}]$$
$$[g] = [\mathcal{L}]/[\mathcal{T}]^2;$$

– *The set of independent dimensionless parameters =*

$$\{v_{y,0}/v_{x,0}\};$$

- *The principal scales—the length scale and the time scale, examples of*:

$$\mathcal{L} = v_{y,0}^2/g$$
$$\mathcal{T} = v_{y,0}/g \, ;$$

- *Solution for the unknowns*:

$$[L_x] = [\mathcal{L}] \Rightarrow L_x \sim (v_{y,0}^2/g) \times \Phi(v_{y,0}/v_{x,0}) \, .$$

The problem is overdetermined:

$$\boxed{L_x \sim (v_{y,0}^2/g) \times \Phi(v_{y,0}/v_{x,0})}$$

where $\Phi(\xi)$ is any dimensionless function.

(b) *Now, assume that x and y are measured in* different *units and solve the problem.*

Solution:

- *The principal units—the units of* horizontal *length, the units of* vertical *length, and the units of time*:

$$[\mathcal{L}_x] \, , \quad [\mathcal{L}_x] \, , \quad [\mathcal{T}] \, ;$$

- *The input parameters and their units*:

$$[v_{x,0}] = [\mathcal{L}_x] / [\mathcal{T}]$$
$$[v_{y,0}] = [\mathcal{L}_y] / [\mathcal{T}]$$
$$[g] = [\mathcal{L}_y] / [\mathcal{T}]^2 \, ;$$

- *The set of independent dimensionless parameters* $= \emptyset$
- *The principal scales— the horizontal length scale, the vertical length scale, and the time scale, examples of*:

$$\mathcal{L}_x = v_{x,0} v_{y,0}/g$$
$$\mathcal{L}_y = v_{y,0}^2/g$$
$$\mathcal{T} = v_{y,0}/g \, ;$$

- *Solution for the unknowns*:

$$[L_x] = [\mathcal{L}_x] \Rightarrow L_x \sim v_{x,0} v_{y,0}/g \, .$$

Finally,

$$\boxed{L_x \sim v_{x,0} v_{y,0}/g}$$

11.1.5   *Kadomtsev-Petviashvili equation*

Consider the Kadomtsev-Petviashvili equation,

$$\frac{\partial}{\partial x}\left(\frac{\partial}{\partial t}u + \left(\frac{\partial^3}{\partial x^3}u + 6u\frac{\partial}{\partial x}u\right)\right) = \frac{\partial^2}{\partial y^2}u\,, \qquad (11.6)$$

where $u = u(x,\,y,\,t)$. It is known to admit solitonic solutions,

$$u(x,\,y,\,t) = \tilde{u}((x - vt)/\Delta x,\, y/\Delta y)\,.$$

Using dimensional analysis, prove that the solitons form families where the
x− and y− widths scale inversely proportionally to the square root of the
velocity and inversely proportionally to the velocity itself, respectively:

$$\Delta x \propto \frac{1}{\sqrt{v}}$$

$$\Delta y \propto \frac{1}{v}\,.$$

Show also that the height of a soliton is proportional to its velocity:

$$\tilde{u}(0,\,0) \propto v\,.$$

<u>Solution:</u> Introduce a constant η such that

$$\frac{\partial}{\partial x}\left(\frac{\partial}{\partial t}u + 6u\frac{\partial}{\partial x}u + \eta\frac{\partial^3}{\partial x^3}u\right) = \frac{\partial^2}{\partial y^2}u\,,$$

   – *The principal units—the units of horizontal (i.e. along X) length*
   and the units of time:

$$[\mathcal{L}_x]\,, \quad [\mathcal{T}]\,.$$

As we will see below

    (a) the vertical coordinate (along Y) will be measured in units
   different from the ones used to measure x. The projectile
   Problem 11.1.4 shows that in some cases, such separation may
   reduce the number of independent dimensionless parameters;
   (b) unlike in Problem 11.1.4, in our case, the units for the
   vertical coordinate are going to be *derived* units;

   – *The input parameters and their units:*

$$[\eta] = [\mathcal{L}_x]^3 / [\mathcal{T}]$$
$$[v] = [\mathcal{L}_x] / [\mathcal{T}]\,;$$

   – *The set of independent dimensionless parameters* $= \emptyset\,;$

– *The principal scales—the horizontal length scale and the time scale, examples of*:

$$\mathcal{L}_x = \left(\frac{\eta}{v}\right)^{1/2}$$

$$\mathcal{T} = \left(\frac{\eta}{v^3}\right)^{1/2};$$

– *A derived scale—the vertical length scale, example of*:

$$\mathcal{L}_y = \mathcal{L}_x^3/\mathcal{T} = \frac{\sqrt{\eta}}{v}.$$

Accordingly, the vertical coordinate y is measured in the units of $[\mathcal{L}_x]^3/[\mathcal{T}]$;

– *Solution for the unknowns*:

$$[\Delta x] = [\mathcal{L}_x] \Rightarrow \Delta x \sim \mathcal{L}_x = \left(\frac{\eta}{v}\right)^{1/2}$$

$$[\Delta y] = [\mathcal{L}_y] = \mathcal{L}_x^3/\mathcal{T} \Rightarrow \Delta y \sim \mathcal{L}_y = \frac{\sqrt{\eta}}{v}$$

$$[\tilde{u}(0,0)] = [\mathcal{L}]/[\mathcal{T}] \Rightarrow \tilde{u}(0,0) \sim \mathcal{L}/\mathcal{T} = v.$$

We can now return to the original dimensionless form of the Eq. (11.6); setting $\eta \to 1$ we get

$$\Delta x \propto \frac{1}{\sqrt{v}}$$
$$\Delta y \propto \frac{1}{v}$$
$$\tilde{u}(0,0) \propto v$$

Remark: The exact Kadomtsev-Petviashvili soliton reads

$$u(x,y,t) = 4A(v)\frac{-\left(\dfrac{x-vt}{\Delta x(v)}\right)^2 + 2\left(\dfrac{y}{\Delta y(v)}\right)^2 + \dfrac{3}{2}}{\left(\left(\dfrac{x-vt}{\Delta x(v)}\right)^2 + 2\left(\dfrac{y}{\Delta y(v)}\right)^2 + \dfrac{3}{2}\right)^2},$$

where $A(v) = v/2$, $\Delta x(v) = \sqrt{2/v}$, and $\Delta y(v) = 2/v$.

11.1.6 *The nonlinear transport equation*

Consider a nonlinear transport equation

$$\frac{\partial}{\partial t}u + u\frac{\partial}{\partial x}u = 0\,. \qquad (11.7)$$

It describes, in particular, the time evolution of a velocity field for a one-dimensional gas of particles. It is assumed that initially, position of an atom defines its velocity uniquely. However, as time develops, faster particles may overcome the slow ones: in this case, three (or more) velocities per position emerge. Continuity of functions would require that starting from this moment, where will be a point in space where the spatial derivative of the density diverges (see Fig. 11.1), and consequently, an (integrable) divergence in density—a *shock wave*—appears (see Fig. 11.2). Later, more shock waves may form. Here and below, we will be considering an initial condition

$$u(x,\,0) = \begin{cases} v_0\left(1 - \left(\frac{x}{a}\right)^2\right) & \text{for } -a \le x \le +a \\ 0 & \text{otherwise} \end{cases}. \qquad (11.8)$$

Using dimensional analysis, ...

   (a) *...prove that the first shock wave will form at $t_{shock\ wave}$ that can be estimated as*

$$t_{shock\ wave} \sim \frac{a}{v_0}\,;$$

Solution: Notice that the nonlinear transport equation (11.8) is dimensionally consistent if the field u is treated, as intended, as a velocity field, and x and t are coordinate and time respectively. Below, we follow the procedure described in Section 1.1.1.

   – *The principal units—the units of length and the units of energy:*

$$[\mathcal{L}]\,, \quad [\mathcal{T}]\,;$$

   – *The input parameters and their units:*

$$[v_0] = [\mathcal{L}]\,[\mathcal{T}]^{-1}$$
$$[a] = [\mathcal{L}]\,;$$

   – *The set of independent dimensionless parameters* $= \emptyset$. Indeed, assume there exists a dimensionless parameter P, that can be expressed through the input:

$$P = \text{const} \times v_0^{\xi_1} a^{\xi_2}\,.$$

Its units are now

$$[P] = [v_0]^{\xi_1}[a]^{\xi_2}$$
$$= [\mathcal{L}]^{\xi_1+\xi_2}[\mathcal{T}]^{-\xi_1}.$$

The analogue the matrix $\hat{M}$ in (1.4) is

$$\hat{M} = \begin{pmatrix} 1 & 1 \\ -1 & 0 \end{pmatrix}.$$

If the problem did not have any dimensionless parameters a priori (like in our case), the number of independent dimensionless parameters will be given by the difference between the number of independent dimensionful parameters and the rank of matrix $\hat{M}$. The later and the former are equal in our case, showing that no independent dimensionless parameters are present;

– *The principal scales—the length scale and the energy scale, examples of*:

$$\mathcal{L} = a$$
$$\mathcal{T} = \frac{a}{v_0};$$

– *Solution for the unknown*:

$$[t_{\text{shock wave}}] = [\mathcal{T}] \Rightarrow t_{\text{shock wave}} = \text{const} \times \mathcal{T} = \text{const} \times \frac{a}{v_0}.$$

where const is a number of the order of unity $\boxed{Q.E.D.}$. This estimate is confirmed by the exact value of $t_{\text{shock wave}}$,

$$t_{\text{shock wave}} = \frac{1}{2}\frac{a}{v_0},$$

that follows from the exact solution given below.

Remark: The exact solution, that can be obtained using the so-called method of characteristics, is represented by a time-parametrized family of parabolas on a zero-velocity pedestal:

$$((t^2u^2 - 2tux + x^2 + ua/v_0 - a^2 = 0) \cap (u \geq 0)) \cup ((u = 0) \cap (|x| > a)).$$
$$(11.9)$$

For $t < \frac{1}{2}\frac{a}{v_0}$, the Eq. 11.9 describes a single valued function of x, $u(x,t)$. At $t = \frac{1}{2}\frac{a}{v_0}$, the velocity field $u(x,t)$ develops an infinite slope at $x = a$, and a shock wave is born. For later times, $t > \frac{1}{2}\frac{a}{v_0}$, each spatial point with

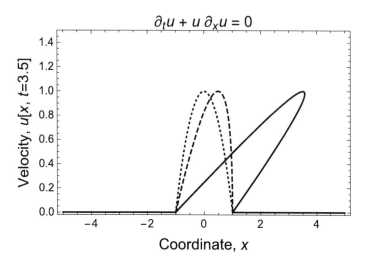

Fig. 11.1   Velocity field governed by a nonlinear transport equation, for a particular initial condition. Initial state (dotted line), velocities at the moment of the formation of the shock wave, $t = 0.5$ in our case (dashed line), velocities at $t = 3.5$ (solid line).

$x > a$, produces three velocities u (see Fig. 11.1). The infinite slope point will persist; it will be moving via a trajectory

$$x_{\text{shock wave}}(t) = v_0 t + \frac{a^2}{4 v_0 t}$$
(11.10)

$$t \geq t_{\text{shock wave}} .$$

It can be shown, using the Lagrange representation of hydrodynamics,[4] that when u is indeed interpreted as a velocity field of a moving gas, its density will diverge at the shock wave point (11.10) (see Fig. 11.2).

   (b) ... *prove that the first shock wave will propagate at a velocity* $v_{shock\ wave}$ *estimated as*

$$v_{shock\ wave} = v_0 .$$

Solution: Solution is almost identical to the one for the sub-problem (a). The only difference is the final conclusion:

   – *Solution for the unknown*:

$$[v_{\text{shock wave}}] = [\mathcal{L}] [\mathcal{T}]^{-1} \Rightarrow v_{\text{shock wave}} = \text{const} \times \frac{\mathcal{L}}{\mathcal{T}} = \text{const} \times v_0 .$$

[4]L.D. Landau and E.M. Lifshitz, *Fluid Mechanics* (Butterworth-Heinemann, Oxford, 1987), 2nd english edition revised.

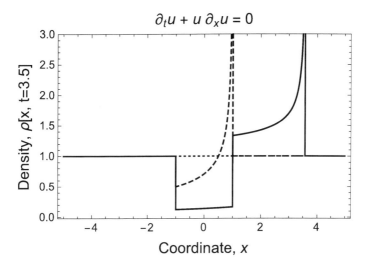

Fig. 11.2 Density distribution for the same initial state and the same instances of time as at Fig. 11.1.

$\boxed{Q.E.D.}$. The exact trajectory (11.10) of the shock wave tip shows that at large times, the shock wave velocity saturates at

$$v_{\text{shock wave}} = v_0 .$$

(c) ... *prove that the total number of the shock waves formed does not depend on the velocity v_0.*

Solution: Any qualitative change in the character of the solution must occur for a particular value of a dimensionless parameter that depends on v_0: if the parameter that governs the number of shock waves were dimensionful, the latter would depend on the choice of units of measurement which is unphysical. On the other hand, we have shown above that in this problem, no dimensionless parameters exist. The only possible choice that remains is that the number of the shock waves is the same for any v_0 $\boxed{Q.E.D.}$. Indeed, the exact solution (11.9) shows that for our class of the initial conditions, one and only one shock wave is supported.

11.1.7 *Burgers equation*

Consider Burgers equation

$$\frac{\partial}{\partial t}u + u\frac{\partial}{\partial x}u = \frac{\partial^2}{\partial x^2}u . \tag{11.11}$$

It can be viewed as a nonlinear transport equation where the shock waves are counterbalanced by viscosity. Indeed, the term in the right-hand-side does tends to decrease the curvature of the velocity curve, and in absence of other effects, force the velocity field to converge to its asymptotic value at infinity. If one neglects the transport term, the second term in the left-hand-side, the equation formally reduces to a diffusion equation. The solution of the Burgers equation, albeit obtainable via elementary methods, is much more complicated than the one for the nonlinear transport equation. Below, we will use a very simple initial state,

$$u(x, 0) = \begin{cases} v_0 & \text{for } -a \leq x \leq +a \\ 0 & \text{otherwise} \end{cases}, \tag{11.12}$$

to make sure the solution fits one page. Remark that unlike for the parabola we used in the transport case, here the shock wave appears right from the start. And it is immediately destroyed by the viscosity (Fig. 11.3).

(a) *For the initial condition (11.12), solve the pure nonlinear transport equation, $\frac{\partial}{\partial t}u + u\frac{\partial}{\partial x}u = 0$. It may help to view the field $u(x, t)$ as a list of coordinates and velocities of freely moving particles.*

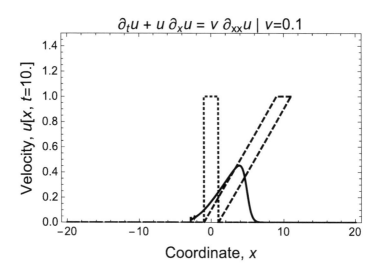

Fig. 11.3   Velocity field in the case of Burgers equation, at $t = 10.$, for a rectangular initial condition. Viscosity coefficient is $\nu = 0.1$ (solid line). Also shown: the initial state (dotted line) and the velocities in the inviscid case, $\nu = 0$, (dashed line), for the same instance of time $t = 10$.

Solution: The resulting velocity field is unique for $x < +a$ and for $x > +a + v_0 t$, and returns three different velocities for $+a \leq x \leq +a + v_0 t$. Namely,

$$u(x, t) = \begin{cases} 0 & \text{for} & x \leq -a \\ \frac{x+a}{t} & \text{for} & -a < x \leq -a + v_0 t \\ \begin{bmatrix} v_0 \\ \frac{x-a}{t} \\ 0 \end{bmatrix} & \text{for} & -a + v_0 t < x \leq +a + v_0 t \\ 0 & \text{for} & x > +a + v_0 t \end{cases}$$

(see the dashed line at Fig. 11.3).

(b) *Show that*

$$u(x, t) = \frac{1}{\sqrt{4\pi t}} e^{-\frac{x^2}{4t}} \tag{11.13}$$

obeys (11.11) if the transport term $u\frac{\partial}{\partial x}u$ is neglected.

Solution: Can be verified by a direct substitution of the solution (11.13) to

$$\frac{\partial}{\partial t} u = \frac{\partial^2}{\partial x^2} u .$$

(c) *Using dimensional analysis, prove that the transition from the shock-wave-domintated to the viscosity-dominated (Fig. 11.4) regime occurs at $av_0 \sim 1$. To do so, supplement the dimensionally meaningless Eqn. 11.11 with relevant constants, to align the dimensionalities of the left-land- and right-hand- sides.*

Solution: The Burgers equation (11.11) as it is written is dimensionally inconsistent if t and x^2 are measured in different units. The way to restore consistency is to introduce, temporarily, a conversion constant ν of a dimensionality $[\mathcal{L}]^2 / [\mathcal{T}]$:

$$\frac{\partial}{\partial t} u + u\frac{\partial}{\partial x} u = \nu\frac{\partial^2}{\partial x^2} u . \tag{11.14}$$

For the field u treated as a velocity field, ν is to be understood as a viscosity coefficient. Let us identify the unique dimensionless parameter that governs the crossover between the shock-wave-dominated and viscousity-dominated regimes.

We follow the procedure described in Section 1.1.1 of BEQM.

– *The principal units—the units of length and the units of energy:*

$$[\mathcal{L}], \quad [\mathcal{T}];$$

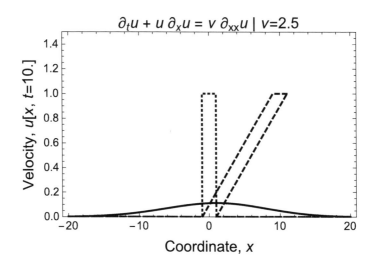

Fig. 11.4   The same as at Fig. 11.3 but with $\nu = 2.5$.

– *The input parameters and their units*:

$$[v_0] = [\mathcal{L}]\,[\mathcal{T}]^{-1}$$
$$[a] = [\mathcal{L}]$$
$$[\nu] = [\mathcal{L}]^2\,[\mathcal{T}]^{-1}\,;$$

– *The set of independent dimensionless parameters* $= \{P \equiv \frac{av_0}{\nu}\}$;
The transition between shock-wave ($P \gg 1$) and an smooth viscosity dominated broadening and damping ($P \ll 1$) should happen at
$P \sim 1$, or at $av_0 \sim \nu$. Returning back to the original convention
$\nu = 1$, we get $av_0 \sim 1$ $\boxed{Q.E.D.}$.

For comparison, the exact solution of (11.11) reads:

$$u(x,\,t) = \left(v_0 e^{-\frac{1}{4}v_0(2x-tv_0-2)}\left(\operatorname{erf}\left(\frac{x-tv_0-1}{2\sqrt{t}}\right) - \operatorname{erf}\left(\frac{x-tv_0+1}{2\sqrt{t}}\right)\right)\right) \Big/$$
$$\left(e^{-\frac{1}{4}v_0(2x-tv_0-2)}\operatorname{erf}\left(\frac{x-tv_0-1}{2\sqrt{t}}\right) - e^{-\frac{1}{4}v_0(2x-tv_0-2)}\operatorname{erf}\left(\frac{x-tv_0+1}{2\sqrt{t}}\right)\right.$$
$$\left. + e^{v_0}\operatorname{erf}\left(\frac{x+1}{2\sqrt{t}}\right) - \operatorname{erf}\left(\frac{x-1}{2\sqrt{t}}\right) - e^{v_0} - 1\right)\,.$$

$$(11.15)$$

11.2 Problems without provided solutions

11.2.1 *Stationary solitons of the Burgers equation*

Consider Burgers equation

$$\frac{\partial}{\partial t}u + u\frac{\partial}{\partial x}u = \frac{\partial^2}{\partial x^2}u\,, \tag{11.16}$$

a toy model for studying questions associated with the celebrated Navier-Stokes equation. *Using dimensional analysis, determine how the size l of a stationary, i.e. obeying*

$$u\frac{\partial}{\partial x}u = \frac{\partial^2}{\partial x^2}u\,,$$

kink solution of this equation depends on its height, $\Delta u > 0$. The presumed structure of the kink soliton is a smoothened step function with

$$u(x) = \begin{cases} +\Delta u\ x \ll -l \\ -\Delta u\ x \gg +l \end{cases};$$

Here, l is the width of the step.

11.2.2 *Stationary solitons of the nonlinear Schrödinger equation*

Consider the nonlinear Schrödinger equation (11.3) again:

$$i\frac{\partial}{\partial t}u = -\frac{\partial^2}{\partial x^2}u - 2|u|^2u$$

$$\int_{-\infty}^{+\infty} dx|u|^2 = N\,;$$

but this time, we are going to use the population of the soliton, N, as an input parameter. It is known that NLSE possesses stationary solitonic solutions (along with the moving solitons that we do not consider here):

$$u(x,\,t) = \frac{u_0}{\cosh(x/\Delta x)}e^{-i\mu t}\,,$$

where the soliton is positioned at $x = 0$, u_0 is the amplitude of the soliton, μ is its chemical potential, and Δx is its width.

The assignment is: *using dimensional analysis only, estimate the width and the chemical potential of the soliton, Δx and μ, as functions of its population N.*

Remark: Again, as in several problems earlier, you have to *invent* suitable units, perform dimensional analysis, and return to the dimensionless form of the equation in the end.

11.2.3   *Solitons of the sine-Gordon equation*

Consider the sine-Gordon equation

$$\frac{\partial^2}{\partial\zeta\partial\eta}u(\zeta,\,\eta) = \sin(u(\zeta,\,\eta))\,. \tag{11.17}$$

In particular, it describes the dynamics of an elongated superconducting Josephson junction.

By inventing proper units for the constituents involved and by inserting some dimensionful constants into the equation if necessary, prove that if Eq. (11.17) admits solitonic solutions

$$u(\zeta,\,\eta) = f((\zeta + v\eta)/\ell)\,,$$

then the soliton size L is proportional to the square root of its speed v:

$$L \propto \sqrt{v}\,.$$

Chapter 12

Rare and exotic methods in elementary quantum mechanics and beyond

12.1 Solved problems

12.1.1 Quantum-mechanical supersymmetry (QM-SUSY): Pöschl-Teller as an example

A good general description of supersymmetric connections in elementary quantum mechanics (QM-SUSY) can be found in Sukumar.[1] In this problem we narrowly focus on a very particular case of a single supersymmetric pair of Hamiltonians, with one of the two members being a free-space Hamiltonian. Furthermore, we look at a single particular example: the Pöschl-Teller potential.

Consider the Schrödinger equation for a one-dimensional particle in a Pöschl-Teller potential,

$$V(x) = -(\hbar^2/ml^2)\operatorname{sech}^2(x/l) \; :$$

$$-\frac{1}{2}\frac{\partial^2}{\partial x^2}\psi - \left(\operatorname{sech}^2 x\right)\psi = E\psi \,,$$

where we set $\hbar = m = l = 1$.

The Pöschl-Teller Hamiltonian,

$$\hat{H} = -\frac{1}{2}\frac{\partial^2}{\partial x^2} - \operatorname{sech}^2 x \,, \tag{12.1}$$

can be factorized as

$$\hat{H} = \hat{A}^\dagger \hat{A} + E_0 \,, \tag{12.2}$$

---

[1] C. V. Sukumar, Supersymmetry, factorisation of the Schrödinger equation and a Hamiltonian hierarchy, *Journal of Physics A* **18**, L57 (1985).

with

$$\hat{A} = \frac{1}{\sqrt{2}} \left(+\frac{\partial}{\partial x} + \tanh x \right) \qquad (12.3)$$

$$\hat{A}^{\dagger} = \frac{1}{\sqrt{2}} \left(-\frac{\partial}{\partial x} + \tanh x \right) \qquad (12.4)$$

$$E_0 = -\frac{1}{2}, \qquad (12.5)$$

Such factorization is at the heart of supersymmetry.

The supersymmetric partner of $\hat{H}$,

$$\hat{H}' = \hat{A}\hat{A}^{\dagger} + E_0,$$

is the free-space Hamiltonian:

$$\hat{H}' = -\frac{1}{2}\frac{\partial^2}{\partial x^2}. \qquad (12.6)$$

Let $\psi(x)$ be an eigenstate of (12.1), with an eigenenergy E. It is easy to show that unless $\hat{A}\psi(x) = 0$, $\psi'(x) \propto \hat{A}\psi(x)$ is an eigenstate of (12.6), of the same energy. Incidentally, when $\hat{A}\psi(x) = 0$, $\psi(x)$ turns out to be an eigenstate of (12.1), typically the ground state.

Likewise, if some other $\psi'(x)$ is an eigenstate of (12.6), then $\psi(x) \propto \hat{A}^{\dagger}\psi'(x)$ is an eigenstate of (12.1), again preserving the energy. In general, we must exclude the possibility $\hat{A}^{\dagger}\psi'(x) = 0$; in our case, this never happens.

(a) *Show that $\psi(x) \propto \tanh x$ is a separatrix—an $E = 0$ eigenstate separating the bound and unbound states—of the Pöschl-Teller Hamiltonian.* To this end, show that it is a supersymmetric image of the $E = 0$ eigenstate of the free Hamiltonian.

Solution: Indeed, at $E = 0$, $\psi'(x) = 1$: hence

$$\psi(x) \propto \hat{A}^{\dagger}\psi'(x) = \frac{1}{\sqrt{2}}\tanh x.$$

(b) *Show that $\psi(x) \propto \operatorname{sech} x$ is a bound state (the ground state, in fact) of the Pöschl-Teller well. Find its energy.* To accomplish this goal, show that it is in the kernel of the "annihilation operator" $\hat{A}$ (i.e. that $\hat{A}\psi(x) = 0$).

Solution: Indeed,

$$\hat{A}\psi(x) = \frac{1}{\sqrt{2}} \left(+\frac{\partial}{\partial x} + \tanh x \right) \operatorname{sech} x = 0.$$

12.1.2 What the supersymmetric structure alone implies for the scattering states of the Pöschl-Teller potential

Consider a generic quantum one-dimensional Hamiltonian with a localized potential,

$$\hat{H} = -\frac{1}{2}\frac{\partial^2}{\partial x^2} + V(x) \tag{12.7}$$

$$V(x) \overset{x \to \pm\infty}{\to} 0 \,.$$

Generically, the scattering eigenstates (i.e. the $E \geq 0$ eigenstates)

$$\psi(x) \approx \begin{cases} \exp[+ikx] + r\exp[ikx] & x \to +\infty \\ t\exp[+ikx] & x \to -\infty \end{cases} \tag{12.8}$$

$$\hat{H}\psi(x) = E\psi(x) \tag{12.9}$$

$$E = \frac{\hbar^2 k^2}{2m} \,, \tag{12.10}$$

where $\hbar k$ is the momentum of the incident wave, and r and t are the reflection and the transmission coefficients, respectively, bound by the unitarity requirement $|t|^2 = 1 - |r|^2$.

Imagine now that one was made aware of the existence of a supersymmetric connection between $\hat{H}$ and the free space, but the details of the supersymmetric map remained unknown. That is

$$\hat{H} = \hat{A}^\dagger \hat{A} + E_0$$

$$\hat{H}' \equiv \hat{A}\hat{A}^\dagger + E_0 = -\frac{1}{2}\frac{\partial^2}{\partial x^2} \,,$$

with

$$\hat{A} = \frac{1}{\sqrt{2}}\left(+\frac{\partial}{\partial x} + w(x) \right)$$

$$\hat{A}^\dagger = \frac{1}{\sqrt{2}}\left(-\frac{\partial}{\partial x} + w(x) \right)$$

$$w(x) \approx \begin{cases} +\sqrt{-E_0} & x \to +\infty \\ -\sqrt{-E_0} & x \to -\infty \end{cases} \,,$$

where

$$w(x), \, E_0 = \text{unknown} \,.$$

Given the uncertainties, what can one still say about the structure of the scattering states?

Hint 1: Recall that if some $\psi'(x)$ is an eigenstate of $\hat{H}'$ of energy E, then

$$\psi(x) \propto \hat{A}^{\dagger} \psi'(x)$$

is an eigenstate of $\hat{H}$, with the same energy E.

Hint 2: Observe that $\hat{A}$ and $\hat{A}^{\dagger}$ are linear differential operators with *asymptotically constant* coefficients.

<u>Solution:</u> No solution.

<u>Corollary:</u> For a generic Hamiltonian $\hat{H}$, its separatrices $\psi(x)$ (i.e. $E = 0$ eigenstates) diverge linearly at large distances:

$$\psi(x) \overset{x \to +\infty}{\propto} x$$

or

$$\psi(x) \overset{x \to -\infty}{\propto} x \, .$$

However, in our case, one of the separatrices remains finite:
$$\psi(x) \overset{x \to \pm\infty}{\propto} \text{const.}$$

12.1.3   *Power-index method. Example of the nonlinear Schrödinger equation, with the Pöschl-Teller problem as a byproduct*

The power-index method introduced by Abraham-Shrauner[2] is based on the observation that for certain finite sets of functions, if one takes a polynomial in functions from this set and differentiates it, the result is another polynomial in functions from the same set. (Mathematicians describe this by saying that "the field of polynomials in functions from this set is closed under differentiation".) The method was developed to generate exact solutions to nonlinear PDEs. In what follows, we will adapt the ideas behind this method to nonlinear ODEs. We will then use the results to generate potentials such that, when these potentials appear inside the linear Schrödinger equation, the latter has solutions that are expressible through elementary functions.

In its simplest form, the power index method deals with the field of polynomials in the function $\tanh x$, which can be shown to be closed under differentiation. Indeed one observes that

$$\frac{d}{dx} \sum_{n=0}^{N} a_n \tanh^n x = - \sum_{n=1}^{N} a_n n (\tanh^{n+1} x - \tanh^{n-1} x) \, .$$

---

[2]B. Abraham-Shrauner, Exact solutions of nonlinear partial differential equations, *Discrete and Continuous Dynamical Systems - Series S* **11**, 577 (2018).

In order for an ODE to have a solution of the form

$$\psi(x) = \sum_{n=0}^{N} a_n \tanh^n x,\tag{12.11}$$

in that ODE there must be a way to cancel the dominant order in $\tanh x$.

Consider the stationary self-defocusing nonlinear Schrödinger equation:

$$-\frac{1}{2}\psi'' + \psi^3 = \mu\psi,\tag{12.12}$$

where the solution $\psi(x)$ is supposed to be real; the chemical potential μ is an adjustable parameter. Assume that there exists a solution of (12.12) in the form (12.11). The three terms in the equation produce the following dominant powers of $\tanh x$:

$$
\begin{array}{ccc}
-\dfrac{1}{2}\psi'' & & \tanh^{N+2} x \\[2mm]
\psi^3 & & \tanh^{3N} x \\[2mm]
\mu\psi & & \tanh^N x\,.
\end{array}\tag{12.13}
$$

This leads to

$$N + 2 = 3N\,,$$

or

$$N = 1\,.$$

Thus if a solution of the form (12.11) exists, it must have the form

$$\psi(x) = a_0 + a_1 \tanh x\,.$$

Substituting this expression into (12.12) produces the "dark soliton" of the NLSE:

$$\psi(x) = \tanh x\,,$$

with $\mu = 1$. The very same solution, $\psi(x) = \tanh x$, can be now reinterpreted as a separatrix—a state separating the bound and unbound states—of the Pöschl-Teller well:

$$-\frac{1}{2}\psi'' - \mathrm{sech}^2(x)\psi = E\psi$$

$$E = 0\,.$$

Above, we reinterpreted $\psi^3(x)$ of (12.12) as $(V(x) + \mu)\psi(x)$, with $V(x) = \psi^2(x) - \mu = -\mathrm{sech}^2(x)$.

Consider now another expansion:

$$\psi(x) = \sum_{n=0}^{N} a_n \operatorname{sech}(x) \tanh^n(x) . \tag{12.14}$$

Observe that

$$\frac{d}{dx} \operatorname{sech}(x) \sum_{n=0}^{N} a_n \tanh^n(x) =$$

$$- \operatorname{sech}(x) \sum_{n=0}^{N} a_n ((n+1) \tanh^{n+1}(x) - n \tanh^{n-1}(x)) .$$

Show that among the solutions of the stationary self-focusing *nonlinear Schrödinger equation*

$$-\frac{1}{2} \psi'' - \psi^3 = \mu \psi , \tag{12.15}$$

the solution

$$\psi(x) = \operatorname{sech} x \; with \; \mu = -\frac{1}{2}$$

is the only one that has the form (12.14).

Solution: The analogue of the table (12.13) produces the following dominant powers:

$$-\frac{1}{2}\psi'' \qquad \operatorname{sech} x \tanh^{N+2} x$$

$$\psi^3 \qquad \operatorname{sech} x \tanh^{3N+2} x$$

$$\mu\psi \qquad \operatorname{sech} x \tanh^{N} x .$$

This leads to

$$N + 2 = 3N + 2 ,$$

or

$$N = 0 .$$

Thus if a solution of (12.15) of a form (12.14) exists, it must have the form

$$\psi(x) = a_0 \operatorname{sech} x .$$

Substitution of this expression to (12.15) produces a "bright soliton" of the NLSE:

$$\psi(x) = \operatorname{sech}(x) ,$$

with $\mu = -1/2$. Again, the very same solution, $\psi(x) = \operatorname{sech} x$, can be now reinterpreted as a bound state (the ground state, in fact) of the Pöschl-Teller well:

$$-\frac{1}{2}\psi'' - \left(\operatorname{sech}^2 x\right)\psi = E\psi \qquad (12.16)$$

$$E = -\frac{1}{2}\,.$$

Above, we reinterpreted $\psi^3(x)$ of (12.15) as $-V(x)\psi(x)$, with $V(x) = -\psi^2(x) = -\operatorname{sech}^2 x$.

12.1.4   *A stationary-kink solution of the Burgers equation through the power-index method*

Consider the stationary version of the Burgers equation (11.11):

$$u\frac{\partial}{\partial x}u = \frac{\partial^2}{\partial x^2}u\,. \qquad (12.17)$$

Using the power-index method, version (12.11), find a kink solution of this equation (see Problem 12.1.3).

   Answer:

$$u(x) = -2\tanh(x)$$

12.1.5   *Scale invariance: quantum Calogero potential as an example*

Consider a three-dimensional quantum particle in a repulsive Calogero potential:[3]

$$\hat{H} = -\frac{\hbar^2}{2m}\Delta + \frac{g}{r^2}\,, \qquad (12.18)$$

where $\Delta \equiv \nabla^2$ is the three-dimensional Laplacian, m is the particle mass, and g is a coupling constant of any sign. The Schrödinger equation reads

$$\hat{H}\Psi = E\Psi\,.$$

After separating the radial and angular motions,

$$\Psi(r,\,\theta,\,\phi) = \frac{\chi(r)}{r}Y_{l,\,m}(\theta,\,\phi)\,,$$

---

[3]F. Calogero, Ground state of a one-dimensional N-body system, *Journal of Mathematical Physics* **10**, 2197 (1969).

the Schrödinger equation for the radial motion becomes

$$-\frac{\hbar^2}{2m}\frac{\partial^2}{\partial r^2}\chi + \frac{\tilde{g}}{r^2}\chi = E\chi, \qquad (12.19)$$

with

$$\tilde{g} \equiv g + \frac{\hbar^2 l(l+1)}{2m}.$$

Recall that the $Y_{l,m}(\theta, \phi)$ are the spherical harmonics, the eigenfunctions of the angular part of the Laplacian. The square of the angular momentum, L^2, and the projection of the angular momentum onto the z-axis read

$$L^2 = \hbar^2 l(l+1)$$
$$M = \hbar m,$$

with

$$l = 0, 1, 2, \ldots, \infty$$
$$m = 0, \pm 1, \pm 2, \ldots, \pm l.$$

Also recall that $L^2/(2mr^2)$ is the centrifugal potential. The radial motion wavefunction, $\chi(r)$, is normalized in such a way that its square gives the radial density. Thus, the units in which $\chi(r)$ is measured are $1/\sqrt{length}$.

Let us concentrate on the case

$$E = 0,$$

the so-called *separatrix*, which separates the bounded and unbounded motions. The classical version of this problem, with an attractive potential,[4] has been first considered in Newton's *Principia*.[5] The resulting Schrödinger equation reads

$$-\frac{\hbar^2}{2m}\frac{\partial^2}{\partial r^2}\chi + \frac{\tilde{g}}{r^2}\chi = 0. \qquad (12.20)$$

(a) Observe a remarkable property of the Calogero potential, the so-called *scale invariance*: the constant $\tilde{g}$ is measured in the same units,

---

[4] ..., rather, the inverse version of it: given a trajectory—a logarithmic spiral in this case—find the potential that generates it, ...

[5] I. Newton, *The Mathematical Principles of Natural Philosophy*, (Daniel Adee, New York, 1846), Book 1, Proposition 9, Problem 4. Beyond the trajectory shape, the time dependence of the coordinate can be extracted as a limiting case of Book 1, Proposition 8, Problem 3.

$mass \times length^4/time^2$, as the constant $\hbar^2/2m$. As a result, the potential $-\frac{\beta}{r^2}$ introduces no new dimensionful constants, only a dimensionless number

$$\eta = \frac{\tilde{g}}{\left(\frac{\hbar^2}{2m}\right)} .$$

In the case of the separatrix (12.20), $\hbar^2/m$ becomes the only dimensionful input.

In terms of η, the equation (12.20) becomes

$$\frac{\partial^2}{\partial r^2}\chi = \frac{\eta}{r^2}\chi . \qquad (12.21)$$

What we have here is a coordinate-dependent (if r were time, then "time-dependent") linear second-order ordinary differential equation. Normally, those can not be solved in terms of elementary functions; special functions are invented and studied only in selected, practically important cases. However, in the Calogero case, the equation (12.21) can be solved in elementary functions.

Also note that $\hbar^2/m$ is useless for constructing a solution of (12.20). First of all, it disappears from (12.21). Second of all, even if it didn't disappear, its dimensional content, i.e. $mass \times length^4/time^2$, contains mass, but the dimensional content of χ, i.e. $1/\sqrt{length}$, doesn't. Since we do not have other input constants that contain mass, there is nothing that can cancel mass from the expression for χ: thus the $\hbar^2/m$ must not be present in an expression for χ. This leaves us with

$$\chi(r) = Af(r), \qquad (12.22)$$

where $f(\cdot)$ is such that

$f(\text{dimensinless\_number}) = \text{another\_dimensionless\_number} .$

In other words, $f(r)$ contains no dimensionful constants. Note that the radial coordinate r is still measured in units of length and $\chi(r)$ is still measured in $1/\sqrt{length}$.

Assignment, finally: Using the information in (12.22) and the text below it alone, narrow down the choice for a possible functional form of a solution of (12.20) (or (12.21)), without resorting to the equation itself.

Solution: At first glance, we want the impossible: how can a dimensionless function $f(\cdot)$ of a dimensionful variable r be dimensionally correct? The only way it can is when $f(\cdot)$ is a power law:

$$f(r) \propto r^\beta .$$

The required dimensional content of χ can then be reached by requiring that

$$A \text{ is measured in } \frac{1}{length^{\beta+\frac{1}{2}}} \,.$$

Thus,

$$\boxed{f(r) =\propto r^\beta} \,. \tag{12.23}$$

(b) *The functional form you identified in point (a) is sufficient to generate both linearly independent solutions of* (12.20) *(or* (12.21)*). Find them.*

<u>Solution:</u> When we substitute our result (12.23) into Eq. (12.21), both the left-hand side and the right-hand side turn out to be proportional to $r^{\beta-2}$. Adjusting the constant β leads to the equation

$$\beta(\beta - 1) = \eta \,.$$

It has two solutions, $\beta = \frac{1}{2}(1 \pm \sqrt{1 + 4\eta})$. Oddly, $\eta = 1$ (i.e. $\tilde{g} = \frac{\hbar^2}{2m}$) leads to $\beta = \phi$, ϕ^{-1}, with $\phi = \frac{1}{2}(1 \pm \sqrt{5})$ being the Golden Ratio, but it is not clear if this fact has any physical significance.

Thus,

$$\boxed{\chi(r) = A_+ r^{\frac{1}{2}(1+\sqrt{1+4\eta})} + A_- r^{\frac{1}{2}(1-\sqrt{1+4\eta})}} \,,$$

with, again, $\eta = \tilde{g}/\left(\frac{\hbar^2}{2m}\right)$. Notice that for $\eta < -\frac{1}{4}$, both solutions above become proportional to $\sqrt{r}$ times an r-dependent phase factor:

$$\chi(r) = A_+ \sqrt{r} e^{+i\frac{1}{2}\sqrt{|1+4\eta|}\,\ln(r)} + A_- \sqrt{r} e^{-i\frac{1}{2}\sqrt{|1+4\eta|}\,\ln(r)} \,. \tag{12.24}$$

12.1.6 *Classical Calogero potential: a posteriori manifestations of the scale invariance*

Manifestations of the scale invariance in classical mechanics are usually subtle, and the most drastic effects—such as the isochronicity of the oscillations—require an additional harmonic trap to be added.[6]

---

[6]L. P. Pitaevskii and A. Rosch, Breathing modes and hidden symmetry of trapped atoms in two dimensions, *Physical Review A* **55**, R853 (1997).

Nonetheless, consider a three-dimensional classical particle in an attractive Calogero potential $V(r) = -g/r^2$,

$$H = \frac{p^2}{2m} + \frac{g}{r^2} , \tag{12.25}$$

where m is the particle mass and $g < 0$ is a coupling constant.[7] The Hamiltonian for the radial motion reads

$$H_{\mathrm{r}} = \frac{p_r^2}{2m} + \frac{\tilde{g}}{r^2} ,$$

where

$$\tilde{g} \equiv g + g_L ,$$

$$g_L \equiv \frac{L^2}{2m} ,$$

and L is the magnitude of the angular momentum vector, a conserved quantity in our case.

Let us focus on the case

$$E = 0 ,$$

the so-called *separatrix* that separates the bounded and unbounded motions. In this problem, we will focus on the case

$$g < -g_L$$

or

$$\tilde{g} < 0 .$$

The resulting Newton's equation reads

$$\ddot{r} = -\frac{2|\tilde{g}|}{mr^3} . \tag{12.26}$$

In polar coordinates, in the orbital plane, the following is a solution that corresponds to a particle that starts, at $t = 0$, at the origin and that reaches the polar angle ϕ_0 at time $t_0 > 0$:

$$r(t) = 2^{\frac{3}{4}} \left(\frac{|\tilde{g}|}{m} \right)^{\frac{1}{4}} \sqrt{t} \tag{12.27}$$

$$\phi(t) = \pm \frac{1}{2} \sqrt{\frac{g_L}{|\tilde{g}|}} \ln \left(\frac{t}{t_0} \right) + \phi_0 . \tag{12.28}$$

[7]F. Calogero, Ground state of a one-dimensional N-body system, *Journal of Mathematical Physics* **10**, 2197 (1969).

Notice that the expression for $\phi(t)$ is deliberately overparametrised, being a one-parametric family of functions in reality. One can now extract the orbit, also overparametrised:

$$r(\phi) = r_0 e^{\gamma(\phi - \phi_0)} \qquad (12.29)$$

with

$$\gamma = \pm\frac{1}{2}\sqrt{\frac{|\tilde{g}|}{g_L}},$$

where $r_0 = 2^{\frac{2}{4}}\left(\frac{\tilde{g}}{m}\right)^{\frac{1}{4}}\sqrt{t_0}$ is the distance to the center when $\phi = \phi_0$. The $\pm$ orbits are a pair of logarithmic spirals of opposite chiralities, a fact known already to Newton: in the problem he considered, the spiral appears as a seed for an inverse problem purporting to infer the potential from an orbit.[8]

For completeness, we must add a solution at negative times, $t < 0$:

$$r(t) = 2^{\frac{3}{4}}\left(\frac{|\tilde{g}|}{m}\right)^{\frac{1}{4}}\sqrt{-t}$$

$$\phi(t) = \pm\frac{1}{2}\sqrt{\frac{g_L}{|\tilde{g}|}}\ln\left(\frac{t}{t_0}\right) + \phi_0,$$

where the angle ϕ_0 is reached at $t = t_0 < 0$.

(a) *Show that (12.27) is the only power-law solution (i.e. solution of the form $r(t) = Bt^\beta$) of Eq. (12.26).*

Solution: A straightforward substitution $r(t) \to Bt^\beta$ into (12.26) gives

$$\beta - 2 = -3\beta$$

and

$$\beta(\beta - 1)B = -\frac{2\tilde{g}}{mB^3},$$

leading to (12.27).

(b) Given the separatrix orbit (12.29)—a logarithmic spiral—one can extract the magnitude[9] of the angular momentum from it as

$$L = \sqrt{\frac{2|g|m}{1 + 4\gamma^2}}. \qquad (12.30)$$

---

[8]I. Newton, *The Mathematical Principles of Natural Philosophy*, (Daniel Adee, New York, 1846), Book 1, Proposition 9, Problem 4. Beyond the orbit shape, the time dependence of the coordinate can be inferred from Book 1, Proposition 8, Problem 3.

[9]In addition, the sign of the projection of the angular momentum on the orbital plane can be extracted from the chirality of the orbit, in turn given by the chosen value of the $\pm$ option in (12.29).

Likewise, given a separatrix–a parabola in this case—for a particle interacting with a Coulomb center $V_{Coulomb}(r) = -\alpha/r$:

$$r_{Coulomb}(\phi) = \frac{r_0}{2\cos^2\left(\frac{\phi}{2}\right)},$$

the angular momentum magnitude is

$$L_{Coulomb} = \sqrt{2m\alpha r_0}. \tag{12.31}$$

As we said above, while quantum-mechanically, the scale invariance is clearly manifest and its consequences are apparent, classical manifestations of the scale invariance and their empirical consequences are much more subtle. Nonetheless, *compare the expressions* (12.30) *and* (12.31) *and identify the difference between them, regarding their behavior under a spatial rescaling of the orbit.*

Solution: One observes the following remarkable property: if one attempts to infer the value of the angular momentum from the orbit (12.29), one does *not* need to know the overall scale r_0. Indeed,

$$L = \sqrt{\frac{2|g|m}{1 + 4\gamma^2}}.$$

In other words, if the orbit is distorted through a spatial rescaling by an unknown factor, information about angular momentum will not be lost. This is not the case however for the Coulomb field. There, if all the distances in the orbit are increased by an unknown factor λ, information about angular momentum—which will increase by a factor of $\sqrt{\lambda}$—will be completely lost.

12.1.7 *Classical Calogero potential: a priori manifestations of scale invariance at the Maupertuis-Jacobi level. Finding the zero-energy orbit from symmetries alone*

We have just seen that in quantum mechanics, the scale invariance of the Calogero potential (12.18) is very easy to spot (the coupling constant g and $\hbar^2/m$ are measured in the same units). Surprisingly, in classical mechanics, none of the standard frameworks—Newton's equations, Lagrange, Hamilton, or Hamilton-Jacobi formalisms—manifest the scale invariance a priori. Problem 12.1.6 deals with the a posteriori signs of the scale invariance. Nonetheless, consider the Maupertuis-Jacobi variational principle, which deals exclusively with orbit shapes and gives no information on time dependence.

According to the Maupertuis-Jacobi variational principle, an orbit $\vec{r}(\xi)$ of a particle of mass m with an energy E, moving in a potential $V(\vec{r})$, extremizes the Maupertuis-Jacobi functional

$$\int \sqrt{2m(E - V(\vec{r}))}dl \,, \tag{12.32}$$

where

$$dl = |\frac{d\vec{r}}{d\xi}d\xi| \,,$$

and ξ is a dummy parameter.

   Using the Maupertuis-Jacobi variational principle, prove that the zero-energy orbits of the Calogero potential (12.25), with $g < -\frac{L^2}{2m}$ (L being the magnitude of the angular momentum) are logarithmic spirals (12.29). Do not resort to differential equations.

   Solution: For the zero-energy orbit of the Calogero problem, the Maupertuis-Jacobi functional (12.32) reads

$$\int \sqrt{2m|g|}\frac{dl}{r} \,. \tag{12.33}$$

This functional is manifestly scale-invariant. Indeed, let us take a valid orbit $r(\phi)$ (written in polar coordinates, in the orbital plane) and generate the curve

$$r'(\phi) \equiv \lambda\, r(\phi) \,, \tag{12.34}$$

where λ is an arbitrary real constant. According to the Maupertuis-Jacobi principle, $r'_{\text{orbit}}(\phi)$ is also a valid orbit. Let us show that this property fixes the orbit to a logarithmic spiral.

   First of all, if the transformed orbit (12.34) is a valid zero-energy orbit of (12.34), it has the same angular momentum L as the original orbit $r(\phi)$. Indeed,

   (a) According to (12.34), for a given ϕ, the total momentum $p'(\phi)$ of the transformed orbit will be given by $p'(\phi) = p(\phi)/\lambda$. Indeed, at $E = 0$, the Hamiltonian (12.25) predicts that $p = \sqrt{2m|g|}/r$.
   (b) For a given ϕ, the ratio between the angular (p_ϕ) and radial (p_r) components of the momentum is the same for both orbits.
   (c) By construction, for a given ϕ, $r'(\phi) = \lambda r(\phi)$.

Thus, the projection of the angular momentum on the orbital plane, L_z, has the same value for the transformed and original orbits. Indeed,

$$L'_z = r'(\phi)p'_\phi(\phi) = (\lambda r(\phi))\left(\frac{p_\phi(\phi)}{\lambda}\right) = r(\phi)p_\phi(\phi) = L_z \,.$$

Now, for a given energy and angular momentum, different orbits must be related by a rotation. Hence, we need to require

$$\lambda r(\phi) = r(\phi + \Delta\phi(\lambda)),$$

where $\Delta\phi(\lambda)$ is such that $\Delta\phi(1) = 0$, and is otherwise unknown. We differentiate both sides with respect to λ, and then set $\lambda = 1$. We get

$$r(\phi) = \left(\frac{d}{d\phi}r(\phi)\right)\left(\frac{d}{d\lambda}\Delta\phi(\lambda = 1)\right),$$

so

$$\frac{d}{d\phi}\ln r(\phi) = \frac{1}{\frac{d}{d\lambda}\Delta\phi(\lambda = 1)} \equiv \gamma$$

Notice that the right-hand side above is a constant that does not depend on ϕ. We finally get

$$r(\phi) \propto e^{\gamma\phi},$$

Q.E.D.

12.1.8   *Circle inversion, quantum: zero-energy eigenstates in a $1/r^4$ potential*

The three-dimensional Laplacian,

$$\Delta_{\vec{r}} \equiv \nabla^2 \equiv \frac{\partial^2}{\partial x^2} + \frac{\partial^2}{\partial y^2} + \frac{\partial^2}{\partial z^2},$$

is known to transform under a sphere inversion,

$$\tilde{\vec{r}} = \left(\frac{\mathcal{R}}{r}\right)^2 \vec{r},$$

as follows:

$$\Delta_{\vec{r}}\,\psi[\vec{r}] = \left(\frac{\tilde{r}}{\mathcal{R}}\right)^5 \Delta_{\tilde{\vec{r}}}\,\tilde{\psi}[\tilde{\vec{r}}],$$

with

$$\tilde{\psi}[\tilde{\vec{r}}] = \left(\frac{\mathcal{R}}{\tilde{r}}\right)\psi\left[\left(\frac{\mathcal{R}}{\tilde{r}}\right)^2 \tilde{\vec{r}}\right] \equiv \left(\frac{r}{\mathcal{R}}\right)\psi[\vec{r}].$$

(Do *not* prove it.) Here and below: $\mathcal{R}$ is a constant corresponding to the radius of the inversion sphere; $r = \sqrt{x^2 + y^2 + z^2}$; $\tilde{r} = \sqrt{\tilde{x}^2 + \tilde{y}^2 + \tilde{z}^2}$; $\Delta_{\tilde{\vec{r}}} \equiv \frac{\partial^2}{\partial \tilde{x}^2} + \frac{\partial^2}{\partial \tilde{y}^2} + \frac{\partial^2}{\partial \tilde{z}^2}$.

(a) Prove that if $\psi[\vec{r}]$ is a zero-energy eigenstate of a $1/r^4$ potential well, i.e. if

$$-\frac{\hbar^2}{2m}\Delta_{\vec{r}}\psi[\vec{r}] - \frac{\alpha}{r^4}\psi[\vec{r}] = 0\,, \qquad (12.35)$$

then $\tilde{\psi}[\vec{r}]$, defined above, obeys the potential-free Schrödinger equation

$$-\frac{\hbar^2}{2m}\Delta_{\tilde{r}}\,\tilde{\psi}[\vec{r}] = \tilde{E}\tilde{\psi}[\vec{r}]\,, \qquad (12.36)$$

for some $\tilde{E}$. Find $\tilde{E}$.

Solution: The calculation is long but straightforward. It gives

$$\boxed{\tilde{E} = \frac{\alpha}{\mathcal{R}^4}}$$

Remark: This connection is a particular manifestation of the so-called Bohlin-Arnold-Vassiliev (BAV) map[10]. In general, BAV map relates different potential. However, the $1/r^4$ potential analyzed above is self-dual with respect to the map.

(b) Construct an example of a solution of (12.35). Require that your solution does not diverge at $r \to \infty$. It is allowed, however, to diverge at $r \to 0$, and it is not required to be normalizable. Do *not* verify if the example you found is a solution of (12.35); it is a long calculation.

Solution: Consider

$$\tilde{\psi}(\tilde{\vec{r}}) = \exp[i\tilde{k}\tilde{z}]\,.$$

Notice that the inverse of a circle inversion is a circle inversion. Thus, if $\tilde{\psi}(\tilde{\vec{r}})$ solves (12.36), then

$$\psi[\vec{r}] = \left(\frac{\mathcal{R}}{r}\right)\tilde{\psi}\left[\left(\frac{\mathcal{R}}{r}\right)^2\vec{r}\right]$$

solves (12.35). For our $\tilde{\psi}(\tilde{\vec{r}})$, we get

$$\boxed{\begin{array}{l} \psi(\vec{r}) \propto \dfrac{1}{r}\exp\left[i\,\dfrac{az}{r^2}\right] \\[2mm] a = \sqrt{\dfrac{2m\alpha}{\hbar^2}} \end{array}}$$

Notice that this state does not depend on the inversion circle radius $\mathcal{R}$.

---

[10]V. Arnold, Huygens and Barrow, Newton and Hooke (Birkhäuser, Basel, 1990).

12.1.9   *Self-similar tilings: moment of inertia of an equilateral triangle as a paradigm*

Find the moment of inertia of a filled equilateral triangle of a side length L and mass M, about a vertical axis through its center of mass, without *integrals.*

Solution: A straightforward calculation would involve a 2D integral with non-so-easy-to-deal-with bounds:

$$I_z = M \frac{\int_{\text{side length } L \text{ triangle}} dx dy \, (x^2 + y^2)}{\int_{\text{side length } L \text{ triangle}} dx dy}$$

(see Fig. 12.1).

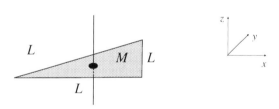

Fig. 12.1   Moment of inertia of an equilateral triangle: the geometry of the problem.

However...

(I)   Observe that on dimensional grounds, the moment of inertia of any filled equilateral triangle of a side length $\mathcal{L}$ and mass $\mathcal{M}$, about a vertical axis through its center of mass, must have the form

$$\mathcal{I}_z = \alpha \mathcal{M} \mathcal{L}^2 \,, \tag{12.37}$$

where α is an unknown constant. Our goal is to find it.

(II) According to the surmise (12.37), the moment of inertia of our triangle of interest is

$$I_z = \alpha M L^2 .\tag{12.38}$$

Let us divide our equilateral triangle onto four smaller equilateral triangles, of a side length $l = L/2$ and mass $m = M/4$ (Fig. 12.2). Observe that according to (12.37), the moment of inertia about the center of mass of each of the smaller triangles is

$$(i_z)_{\mathrm{I}} = (i_z)_{\mathrm{II}} = (i_z)_{\mathrm{III}} = (i_z)_{\mathrm{IV}} = \alpha m l^2 .$$

The moment I_z becomes

$$I_z = (i_z)_{\mathrm{I}} + \left((i_z)_{\mathrm{II}} + m\rho^2\right) + \left((i_z)_{\mathrm{III}} + m\rho^2\right) + \left((i_z)_{\mathrm{IV}} + m\rho^2\right) ,\tag{12.39}$$

where $\rho = \frac{L}{2\sqrt{3}}$ is the distance between the centers of mass of the triangles II, III, and IV from the origin.

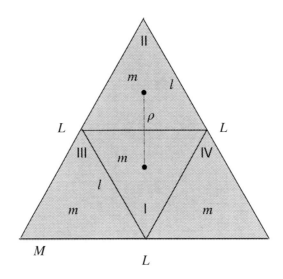

Fig. 12.2   Moment of inertia of an equilateral triangle: a self-similar partition.

(III) Equating the right-hand sides of (12.38) and (12.39), we get

$$\alpha = \alpha \left(\frac{1}{4}\right)\left(\frac{1}{2}\right)^2 + 3\left(\alpha\left(\frac{1}{4}\right)\left(\frac{1}{2}\right)^2 + \left(\frac{1}{4}\right)\left(\frac{1}{2\sqrt{3}}\right)^2\right)$$

or

$$\alpha = \frac{1}{12}.$$

All in all,

$$\boxed{I_z = \frac{1}{12}ML^2}$$

12.1.10   *Properties of the spectrum of an equilateral triangular quantum billiards that follow from the tiling self-similarity*

Consider an equilateral triangular quantum billiards with impenetrable walls, Fig. 12.3. As the figure shows, the billiards allows for a self-similar partition. Imagine that its ground state energy, $E_{g.s.} = \alpha \frac{\hbar^2}{mL^2}$, is known. Here, L is the side length, m is the particle mass, and α is a dimensionless constant. *Using a self-similarity argument, prove that the energy spectrum contains a ladder*

$$E_m = 4^m E_{g.s.} \tag{12.40}$$

with $m = 0, 1, 2, 3, \ldots$

Solution: Let $\psi_{g.s.}(\vec{r})$ be the ground state of the large billiard. Assume that it is non-negative. Consider the ground states of the four smaller billiards, I, II, III, and IV: $\psi_{g.s.,\ I(II,\ III,\ IV)}(\vec{r})$. Assume that all four are non-negative as well. The ground state energy of each of the four smaller billiards is $\epsilon_{g.s.} = \alpha \frac{\hbar^2}{ml^2} = 4E_{g.s.}$, where $l = L/2$ is the side length of a smaller billiards.

Now, construct the state

$$\psi_{m=1}(\vec{r}) = \frac{1}{2}\left(\psi_{g.s.,\ I}(\vec{r}) - \psi_{g.s.,\ II}(\vec{r}) - \psi_{g.s.,\ III}(\vec{r}) - \psi_{g.s.,\ IV}(\vec{r})\right).$$

This state (a) is an eigenstate of the kinetic energy, (b) obeys the boundary conditions of the original problem, and (c) can be shown to have continuous first derivatives at the junctions between the smaller billiards. This means that the state $\psi_{m=1}(\vec{r})$ is a legitimate eigenstate of the large billiards, with an energy $E_{m=1} = 4E_{g.s.}$. The procedure can be then repeated for each of the smaller billiards, leading to an eigenstate with an energy $E_{m=2} =$

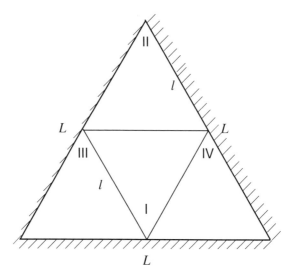

Fig. 12.3   Equilateral trianglular quantum billiards with hard walls and its self-similar partition.

$16E_{\text{g.s.}}$. The process can be repeated ad infinitum leading to (12.40) *Q.E.D.*
<u>Remark</u>: the known exact spectrum of our billiards,

$$E_{n_1, n_2} = \frac{8}{9} \frac{\hbar^2}{mL^2} (n_1^2 + n_1 n_2 + n_2^2) \tag{12.41}$$

$$n_1 = 1, 2, 3 \ldots \tag{12.42}$$

$$n_2 = n_1 + 1,\, n_1 + 2,\, n_1 + 3,\, \ldots, \tag{12.43}$$

does contain the subsequence (12.40). It can be obtained by setting

$$n_1 = 2^m;\ n_2 = 2^{m+1}.$$

.Notice that the ground state is reached for $n_1 = 1$; $n_2 = 2$; its energy is $E_{\text{g.s.}} \frac{56}{9} \frac{\hbar^2}{mL^2}$ $(\alpha = \frac{56}{9})$.

12.1.11   *Center of mass of a filled "golden b" shape, through self-similar tilings*

Fig. 12.4 describes a "golden b" shape, popular in recreational mathematics.[11] The ratios between its dimensions are proportional to the powers

---

[11]https://web.archive.org/web/20230426000449/https://goldenbpuzzles.com/insights/why-is-the-golden-b-golden/

of

$$\eta = 1/\sqrt{\phi}\,,$$

where

$$\phi = \frac{\sqrt{5}+1}{2}$$

is the golden ratio. Recall that the golden ratio between two quantities s and t appears when we require that the ratio between their sum and the larger of the two equals the ratio between the larger and the smaller ones: $\frac{s+t}{s} = \frac{s}{t} = \phi$, with $s > t$. The property used in our problem is that the golden ratio solves

$$\phi^2 - \phi - 1 = 0\,;$$

the inverse of the golden ratio $1/\phi = \frac{\sqrt{5}-1}{2}$ is also a solution of this algebraic equation. It allows for an elegant self-similar partition shown in Fig. 12.5.

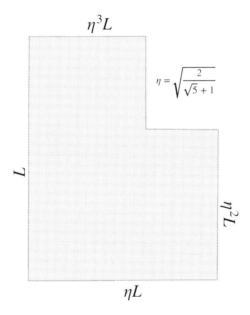

Fig. 12.4   The "golden b": geometry. Here $\eta = 1/\sqrt{\phi}$, with $\phi = \frac{\sqrt{5}+1}{2}$ being the golden ratio.

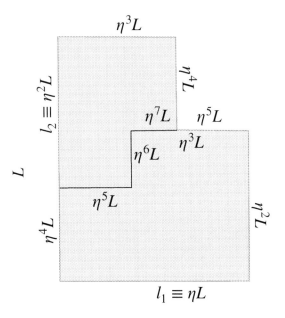

Fig. 12.5   The "golden b": a self-similar partition.

Find the location of the center of mass (distances A and B) of a filled "golden b", without *integrals*. Same for its constituent parts (distances a_1, b_1, a_2, and b_2). See Fig. 12.6.

Solution: The similarity between the smaller "b"'s 1 and 2 and the original "b" implies

$$a_1 = \eta A$$
$$b_1 = \eta B$$
$$m_1 = \eta^2 M$$
$$a_2 = \eta^2 A$$
$$b_2 = \eta^2 B$$
$$m_2 = \eta^4 M \,.$$

On the other hand, the center-of-mass positions are related as

$$A = \frac{m_1(\eta L - b_1) + m_2 a_2}{m_1 + m_2}$$

$$B = \frac{m_1 a_1 + m_2(L - b_2)}{m_1 + m_2}.$$

Here we used the fact that the position of the center of mass of a compound object is a weighted mean of the center-of-mass positions of the constituent parts. We, thus, get

$$A = \frac{\eta L - \eta B + \eta^4 A}{1 + \eta^2}$$

$$B = \frac{\eta A + \eta^2 L - \eta^3 B}{1 + \eta^2}.$$

Solving the system above, we finally get

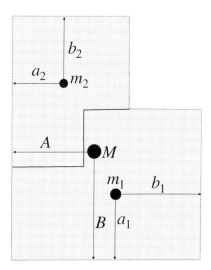

Fig. 12.6   The "golden b": positions of the centers of mass, for the main shape and its constituents. Also, the corresponding masses, M, m_1, and m_2 are indicated.

$$
\begin{aligned}
A &= \frac{\eta L}{\sqrt{5}} \\
B &= \frac{L}{\sqrt{5}} \\
a_1 &= \eta A \\
b_1 &= \eta B \\
a_2 &= \eta^2 A \\
b_2 &= \eta^2 B
\end{aligned}
\qquad (12.44)
$$

12.2 Problems without provided solutions

12.2.1 *Power-index method applied to the Pöschl-Teller problem directly*

Examine the Poshcl-Teller problem (12.16) using the power-index method (see Problem 12.1.3). Provide explicit solutions (if they exist), for both the expansion (12.11) with $N = 0, 1, 2$ and the expansion (12.14) with $N = 0, 1$.

12.2.2 *Quasi-integrability: the overall idea and the example of the sixtic oscillator*

In the context of the quasi-integrability construction of the Background material 12.3.1, consider the particular case $M = 0$ and $p = 0$, and show that in this case, the vacuum state (12.46) is an exact eigenstate of the Hamiltonian (12.45). Find the corresponding eigenvalue $(E_0)_{m=0,\, p=0}$.

Answer:

$$
(E_0)_{m=0,\, p=0} = \frac{\sqrt{3}}{2^{3/4}}
$$

12.2.3   A quasi-integrable instance of two nonlinearly coupled oscillators

Consider a Hamiltonian for two harmonic oscillators, the second having a frequency two times higher than the first, supplemented by a nonlinear perturbation:

$$\hat{H} = \hat{H}_0 + \hat{V} \,,$$

with

$$\hat{H}_0 = \hbar\omega(\hat{n}_1 + \frac{1}{2}) + 2\hbar\omega(\hat{n}_2 + \frac{1}{2})$$

$$\hat{V} = g \left\{ (\hat{a}_1)^2 \hat{a}_2^\dagger + (\hat{a}_1^\dagger)^2 \hat{a}_2 \right\} \,,$$

where $\hat{n}_{1(,2)} \equiv \hat{a}_{1(,2)}^\dagger \hat{a}_{1(,2)}$ is the standard harmonic-oscillator excitation number operator for the first(second) oscillator, and g is the perturbation strength.

(a) Using the first-order (in g) degenerate perturbation theory, find the eigenenergies in the first degenerate manifold of the eigenstates of $\hat{H}_0$,

$$E^{(0)}_{n_1 + 2n_2 = 2} = \frac{5}{2}\hbar\omega \,.$$

Recall that this manifold is two-fold degenerate; it is spanned by the eigenstates $|n_1 = 2, n_2 = 0\rangle$ and $|n_1 = 0, n_2 = 1\rangle$.

(b) A remarkable property of the perturbation $\hat{V}$ is that it commutes with the unperturbed Hamiltonian $\hat{H}_0$. Do *not* prove that.

However, prove that if

$$[\hat{V}, \hat{H}_0] = 0 \,,$$

then $\hat{V}$, acting on an eigenstate of $\hat{H}_0$ of energy $E^{(0)}$, returns a (generally different) eigenstate of $\hat{H}_0$ of the same energy $E^{(0)}$.

(c) Prove that according to (b), the eigenenergies you obtained in (a) are, in fact, *exact*.

Recall that for a single harmonic oscillator of frequency ω we have:

$$\hat{a}^\dagger = \frac{-i}{\sqrt{2\hbar m\omega}}(\hat{p} + im\omega x)$$

$$\hat{a} = \frac{+i}{\sqrt{2\hbar m\omega}}(\hat{p} - im\omega x)$$

$$\hat{a}^\dagger|n\rangle = \sqrt{n+1}|n+1\rangle$$

$$\hat{a}|n\rangle = \sqrt{n}|n-1\rangle \,,$$

where $|n\rangle$ is the nth eigenstate of a harmonic oscillator of frequency ω and mass m, and

$$n = 0, 1, 2, \ldots \quad .$$

Also,

$$E_n = \hbar\omega\left(n + \frac{1}{2}\right).$$

12.2.4 Moment of inertia of a filled "golden b" shape, through the self-similar tilings

Using the results from Problem 12.1.11 and Problem 12.1.9 as an inspiration, find the moment of inertia of a filled "golden b", without integrals.

Answer:

$$\boxed{I = \frac{\eta^2}{5}\,ML^2}$$

12.3 Background

12.3.1 Quasi-integrability: the overall idea and the example of the sixtic oscillator

Quasi-integrability is a rare phenomenon where the Hamiltonian, for a subspace spanned by a finite number of basis states, stabilizes this subspace, i.e. does not have matrix elements that lead away from the subspace. The phenomenon has been intensely studied by Turbiner[12] and Ushveridze.[13] The best-known instance of quasi-integrability is the sixtic oscillator with a specifically chosen sub-leading (at $x \to \pm\infty$) term:

$$\hat{H} = -\frac{1}{2}\frac{\partial^2}{\partial x^2} + x^6 + 2^{3/4}\sqrt{4M + 2p + 3}\,x^4 \tag{12.45}$$

$$M = 0, 1, 2, \ldots$$

$$p = 0, 1.$$

[12] Alexander V. Turbiner, One-dimensional quasi-exactly solvable Schrödinger equations, *Physics Reports* **642**, 1 (2016).
[13] A. G. Ushveridze, *Quasi-exactly Solvable Models in Quantum Mechanics* (Taylor & Francis Group, New York, 1994).

Let us introduce the operators

$$\hat{a}_+ = x$$

and

$$\hat{a}_- = \frac{\partial}{\partial x} + \sqrt{2}x^3 + 2^{1/4}\sqrt{4M + 2p + 3}\,x\,.$$

They are constructed in such a way that $\hat{a}_-$ and $\hat{a}_+$ obey the harmonic-oscillator commutation relations:

$$[\hat{a}_-, \hat{a}_+] = 1\,.$$

The vacuum of this fictitious harmonic oscillator (i.e. the state $\psi_0(x) \equiv |0\rangle$ that obeys $\hat{a}_-|0\rangle = 0$) does exist:

$$\psi_0(x) = A\exp[-\frac{x^4}{2\sqrt{2}} - \frac{\sqrt{4M + 2p + 3}\,x^2}{2^{3/4}}]\,, \qquad (12.46)$$

with

$$A = (2^{-5/8}e^{M+p/2+3/4}(4M + 2p + 3)^{1/4}K_{1/4}(M + p/2 + 3/4))^{-1/2}\,.$$

Here, $K_n(\xi)$ is the modified Bessel function of the second kind. The raising operator $\hat{a}_+$ can now be used to construct a useful Hilbert space basis:

$$|n\rangle = (n!)^{-1/2}(\hat{a}_+)^n|0\rangle\,. \qquad (12.47)$$

In the basis (12.47), matrix elements of the Hamiltonian (12.45) read[14]

$$H_{n+2,\,n} = -\sqrt{2}(2M + p - n)$$

$$H_{n,\,n} = \frac{(2n + 1)\sqrt{4M + 2p + 3}}{2^{3/4}}$$

$$H_{n-2,\,n} = \frac{n(n - 1)}{2}\,.$$

In is evident that for given M and p, the M-dimensional Hilbert subspace spanned by the basis states

$$|0\rangle, |2\rangle, \ldots |2M\rangle$$

for $p = 0$ or by

$$|1\rangle, |3\rangle, \ldots |2M + 1\rangle$$

for $p = 1$ is closed under the action of the Hamiltonian: as a result, one can find M *exact* eigenstates of $\hat{H}$ by diagonalizing an $M \times M$ matrix.

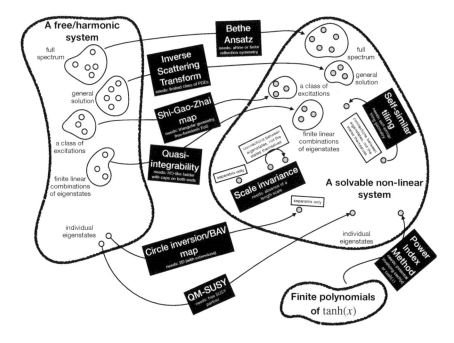

Fig. 12.7

12.3.2 *Relationship between various methods*

The following diagram summarizes the results of Chapter 12. Several remarks are in order:

- The "non-linear" vs. "linear" dichotomy must be understood in the sense of classical mechanics: it refers to the properties of the corresponding classical ODEs/PDEs.
- While Bethe ansatz is discussed below, the subject is barely touched. It is an enormous subject: there are a thousand times more publications on the Bethe ansatz than on any other topic in this chapter. The reason is that unlike the other methods, the

---

[14]Note that the Hamiltonian matrix below is *not* an Hermitian matrix. This is a consequence of the fact that $\hat{a}_-$ and $\hat{a}_+$ are *not* Hermitian conjugates. The conclusions to follow are unaffected by this defect.

Bethe ansatz naturally fits the many-body world and as such counts numerous empirical applications under its belt.[15]

- All methods described in this Chapter have a very limited range of applicability:

  · Bethe ansatz requires an affine or finite reflection symmetry, of which we have fewer than ten.
  · Quasi-integrability requires a fine tuning of the Hamiltonian parameters.
  · Self-similar tiling is a very rare geometric property. In addition, it requires prior knowledge of at least one eigenstate of the non-linear system of interest;
  · Scale invariance is possessed by a very limited group of interactions, including: a Calogero potential, two-dimensional δ-function, infinite strength δ-function. Even then, only the separatrices (zero-energy states) are solvable. At finite energies, a known eigenstate can generate a ladder of eigenstates, not unlike in the self-similar tiling case. The strength of the scale-invariance is that it, similarly to the Bethe ansatz, applies to many-body cases, with rich empirical consequences (see, e.g. Saint-Jalm et al.[16]).
  · The Bohlin-Arnold-Vassiliev map applies to a large variety of two-dimensional potentials supplemented by an inverse quartic potential in any number of spatial dimensions. However, generally, only the separatrix can be solved, being an image of a free problem on the other end. The Coulomb problem is a rare exception, being an image of a harmonic problem.
  · Solvability using quantum-mechanical supersymmetry poses the severe demand for the supersymmetric partner of the system of interest to be a free problem.
  · The power-index method requires the potential to be a finite polynomial of $\tanh x$ and $\operatorname{sech} x$; even then, a solution is not guaranteed.

---

[15]M. T. Batchelor, The importance of being integrable: Out of the paper, into the lab, *International Journal of Modern Physics B* **28,** 1430010 (2014).

[16]R. Saint-Jalm, P. C. M. Castilho, E. Le Cerf, B. Bakkali-Hassani, J.-L. Ville, S. Nascimbene, J. Beugnon and J. Dalibard, Dynamical symmetry and breathers in a two-dimensional Bose gas, *Physical Review X* **9,** 021035 (2019).

- Two other methods did not fit this Chapter:

  · The inverse scattering transform allows solving a limited class of partial differential equations, including the Korteweg-de Vries equation, the nonlinear Schrödinger equation, and the sine-Gordon equation.[17] It maps a nonlinear field of interest to an auxiliary field that follows a linear evolution. The nonlinear field generates a potential with which the linear field interacts.

  · The Shi-Gao-Zhai map solves, for a limited class of initial conditions, time-dependent Thomas-Fermi equations for gases with a free-fermionic equation of state, via a map to a Liouvlille equation for the phase-space density of free fermions.[18]

12.3.3 *Further reading: Bethe ansatz*

The Bethe ansatz[19] is a rich, hundred-year-old field of research, and generally, its applications lie within the realm of many-body physics, far away from the few-body problems discussed here. Mathematically, it deals with either multi-dimensional quantum billiards with various boundary conditions, or wave propagation in a space permeated by the δ-functional semi-transparent mirrors, or a combimation of both. For a particular subclass of these problems, their solutions are finite linear combinations of plane waves. The mathematical phenomenon behind such solvability is an *affine reflection symmetry* the wall and mirror configuration possesses. A purely hard-wall case allows for a concise description.[20]

To give a flavor of what the Bethe ansatz wavefunctions look like, let us list the eigenstates of the triangular billiards of Problem 12.1.10:

$$\psi_{n_1, n_2}(x, y) = \frac{2}{\sqrt{6}} \frac{1}{L} \sum_{\substack{m=0, 1, 2 \\ p=0, 1}} (-1)^p \exp[i\, \hat{\mathcal{R}}_{2m\pi/3}\, \hat{\mathcal{T}}^p\, \vec{k}_{n_1, n_2} \cdot \vec{r}]$$

---

[17]M. J. Ablowitz and H. Segur, *Solitons and the Inverse Scattering Transform* (SIAM, Philadelphia, 1981).

[18]Z.-Y. Shi, C. Gao and H. Zhai, Ideal-gas approach to hydrodynamics *Physical Review X* **11,** 041031 (2021).

[19]M. Gaudin and J.-S. Caux, *The Bethe Wavefunction* (Cambridge University Press, Cambridge, UK, 2014).

[20]M. Olshanii and S. G. Jackson, An exactly solvable quantum four-body problem associated with the symmetries of an octacube, *New Journal of Physics* **17,** 105005 (2015).

$$E_{n_1,n_2} = \frac{8}{9}\frac{\hbar^2}{mL^2}(n_1^2 + n_1 n_2 + n_2^2)$$

$$n_1 = 1, 2, 3 \ldots$$

$$n_2 = n_1 + 1, n_1 + 2, n_1 + 3, \ldots,$$

where

$$k_{n_1,n_2} = n_1 \begin{pmatrix} +\frac{2\pi}{3L} \\ -\frac{2\pi}{\sqrt{3}L} \end{pmatrix} + n_2 \begin{pmatrix} -\frac{2\pi}{3L} \\ -\frac{2\pi}{\sqrt{3}L} \end{pmatrix},$$

$\hat{\mathcal{R}}_\phi$ is a matrix that performs a two-dimensional rotation by an angle ϕ, and $\hat{\mathcal{T}}$ is the reflection matrix about the x-axis.

Further Reading

Hans G. Hornung, *Dimensional Analysis: Examples of the Use of Symmetry*, Dover, Mineola, New York (2006).

Debnarayan Jana, *Dimensional Analysis: Modern Perspectives*, Lambert Academic Publishing, Saarbrücken (2011).

Vladimir P. Krainov, *Qualitative Methods in Physical Kinetics and Hydrodynamics*, American Institute of Physics, College Park (1992).

Don S. Lemons, *A Student's Guide to Dimensional Analysis*, Cambridge University Press, Cambridge (2017).

Harold J. Metcalf, *Topics in Classical Biophysics*, Prentice-Hall, Englewood, New Jersey (1980).

A. B. Migdal, *Qualitative Methods in Quantum Theory*, Westview Press, Boulder (2000).

Clifford Swartz, *Back-of-the-Envelope Physics*, The Johns Hopkins University Press, Baltimore (2003).

Sanjoy Mahajan, *Street-Fighting Mathematics: The Art of Educated Guessing and Opportunistic Problem Solving*, The MIT Press Cambridge-London (2010).

General Index

Author Index